NUMERICAL METHODS FOR SHALLOW-WATER FLOW

Water Science and Technology Library

VOLUME 13

The titles published in this series are listed at the end of this volume.

NUMERICAL METHODS
FOR
SHALLOW-WATER FLOW

by

C. B. VREUGDENHIL

Institute for Marine and Atmospheric Research Utrecht (IMAU),
Utrecht University, Utrecht, The Netherlands

KLUWER ACADEMIC PUBLISHERS

DORDRECHT / BOSTON / LONDON

A C.I.P. Catalogue record for this book is available from the Library of Congress.

ISBN 0-7923-3164-8

Published by Kluwer Academic Publishers,
P.O. Box 17, 3300 AA Dordrecht, The Netherlands.

Kluwer Academic Publishers incorporates
the publishing programmes of
D. Reidel, Martinus Nijhoff, Dr W. Junk and MTP Press.

Sold and distributed in the U.S.A. and Canada
by Kluwer Academic Publishers,
101 Philip Drive, Norwell, MA 02061, U.S.A.

In all other countries, sold and distributed
by Kluwer Academic Publishers Group,
P.O. Box 322, 3300 AH Dordrecht, The Netherlands.

Printed on acid-free paper

Printed in the Netherlands

Contents

Preface

Unlike many other books, this one did not grow out of lecture notes. Rather, I felt that after more than 25 years of widespread application of numerical shallow-water models, a suitable reference book was still missing. Of course, sections on shallow-water flow are included in some CFD books, but until very recently a systematic treatment of both physical and numerical aspects was not available. Most of the literature is scattered over hundreds of papers and reports.

In this text, the theory of shallow-water flow and its numerical simulation is given. I am stressing the physics in the first part of the book because I am convinced that you need to realize which type of solutions to expect. In the second part, typical numerical methods are discussed. This is not a cook-book and I did not include all methods available in literature, but I emphasize how to obtain guidelines for choosing a method for your particular problem, taking such things as stability and, most importantly, accuracy into account.

Most of the book is about two-dimensional flows, the treatment of which is well-established. Recent research, however, is concentrating on 3-d flows. Therefore one chapter has been included which discusses the main aspects of such flows. Case studies are not given. My experience is that published case studies rarely provide sufficient details to be really useful. Moreover, giving case studies for the whole spectrum of applications and the whole spectrum of numerical methods would take far too much space.

In selecting material, I have been slightly biased towards work from The Netherlands. There is some justification for this, as significant contributions have come from this country and from persons closely related with it. Nevertheless, I have attempted to include all relevant contributions, though without any claim to be exhaustive.

The audience of this book is supposed to consist of three (perhaps not disjoint) groups. First of all, I have in mind scientists and engineers involved in applications of shallow-water models; they should have a sufficient knowledge of physics and numerics to judge the reliability of their results. Secondly, developers of fluid-dynamics software should be aware of what has been accomplished and where the problems are. Finally, numerical mathematicians need insight in the type of problems they are developing solution methods for. Although the field of applications is very wide (see chapter 1) and I have tried to do justice to it, the major emphasis is on engineering applications in rivers, estuaries and coastal seas. The book is self-contained but the reader is assumed to have a basic knowledge of fluid mechanics and CFD.

Acknowledgements

I am grateful to dr. B. Koren, dr. J. Versteegh and prof. dr. J.T.F. Zimmerman, who read parts of or the entire manuscript. Their comments led to significant improvements in many places. I also acknowledge advice on several detail questions, references and illustration material from Mr. B. Bosselaar, dr. H. Gerritsen, dr. H. Ridderinkhof and prof. dr. G. S. Stelling.

Chapter 1

Shallow-water flows

1.1. Introduction

Shallow-water flow might seem to be a too special subject to devote a complete book to, but it is actually quite common. Many types of flow, not necessarily involving water as a fluid, can be characterized as shallow-water flows. The general characteristic of such flows is that the vertical dimension is much smaller than any typical horizontal scale and this is true in many everyday situations. This chapter gives a number of illustrations.

Shallow-water flows are nearly horizontal, which allows a considerable simplification in the mathematical formulation and numerical solution by assuming the pressure distribution to be hydrostatic. However, they are not exactly two-dimensional. The flow exhibits a three-dimensional structure due to bottom friction, just as in boundary layers. Moreover, density stratification due to differences in temperature or salinity causes variations in the third (vertical) direction. Yet, in many shallow-water flows these 3-d effects are not essential and it is sufficient to consider the depth-averaged form, which is two-dimensional in the horizontal plane. This restricted form, commonly indicated by the term "shallow-water equations", is the one discussed in this book. It could be argued that it is no longer necessary to make this kind of approximations in the present supercomputer era, where fully 3-d and time-dependent flows can be simulated. However, for many practical applications that is just too much detail and using the 2-d depth-averaged shallow-water equations gives essentially the same information at much lower cost.

The relative simplicity is the reason why such flows have attracted the attention of many mathematicians and hydrodynamicists. This is not a historic review; for more historic information see, e.g. Dronkers (1964). Some names should, however, be mentioned because of terminology. A lot of interesting work on tidal flows was done by Laplace; what is known as the Laplace tidal equations is a somewhat specialized form of the shallow-water equations (see Lamb,1932 or Hendershott, 1981). In the French scientific community, the shallow-water equations are commonly referred to as the Saint-Venant equations, although it appears that Saint-Venant derived only the 1-d version (Dronkers, 1964). In this book, we will use the term shallow-water equations (SWE) throughout.

The numerical solution of the SWE was one of the early applications of digital computers when these became available in the late 1940's. Simulations were done by Charney et al (1950) for atmospheric flows and Hansen (1956) for oceanographic flows. A considerable development has taken place since, up to the present situation where it can be stated that at least the 2-d SWE are well understood. This book attempts to present that understanding, both from a fluid-mechanical (chs. 2 ...5) and a numerical point of view (chs.6...10). The three-dimensional case is treated more briefly in chapter 11.

A few further developments are mentioned:
* Three-dimensional models are getting more and more common, particularly in meteorology (everyday forecasts are produced using sophisticated 3-d models of the shallow-water type).
* In oceanography, as well, 3-d effects are taken into account, either in the form of layered models (with a small number of layers) or using a full 3-d discretization.
* In coastal and river hydrodynamics, 3-d effects are getting increased attention, particularly in view of density stratification.
* Shorter and steeper waves are treated approximately by what is called the Boussinesq equations, which can be considered an extension of the shallow-water equations.
* Very importantly, flow models are extended by transport models for heat, dissolved substances or suspended sediment particles. In the latter case, also the coupling with a mobile bottom can be taken into account, which gives rise to important questions of stability and predictability.

In the following sections, a number of typical applications are briefly described. The examples are not intended to be the most recent or the most elaborate ones, neither to represent all research groups. Rather, it is a personal choice intended to illustrate how wide the field of application is even for the 2-d depth-averaged SWE. The selection is biased towards work done in my immediate surroundings, which does not imply that other similar work might not be equally or perhaps even much more important.

1.2. Atmospheric flows

A good example of the application of the 2-d SWE to atmospheric flows is the pioneering attempt by Charney, Fjörtoft and Von Neumann (1950) to produce a numerical weather forecast. They integrated the hydrodynamic equations across the total depth of the atmosphere, disregarding density stratification, thus arriving at the 2-d barotropic equations. For large-scale atmospheric flows, the Coriolis acceleration and its variation with latitude are essential effects. By making the rigid-lid approximation, fast gravity waves were suppressed, which would be rather a nuisance for the purpose of weather prediction. The dynamic equations were formulated in terms of potential vorticity and stream function.

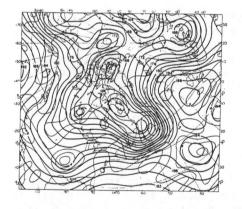 

Fig. 1.1. Forecast of January 30, 1949, 0300 GMT.
Left: observed, right: computed stream function after 24 hours (heavy lines)

Using one of the first electronic computers (the ENIAC), they solved the equations by a combination of finite-difference and spectral techniques on a 15 * 18 numerical grid with a grid size of 736 km. The time step was 1 to 3 hours and the forecast time 24 hours (this, incidentally, took several hours of computing time). Some results are shown in fig. 1.1., illustrating the observed and computed vorticity and stream function before and after the 24 hour period. Although the results were far from perfect, they were considered sufficiently realistic to warrant a further development. This has led to detailed 3-d models used today for everyday weather forecasting, in which many more physical effects are included, such as radiation, cloud formation, precipitation, surface fluxes in land and sea areas etc. Describing this development is outside the scope of this book; see, e.g. Haltiner & Williams (1980).

1.3. Tidal flows

Traditionally, tidal flows have been the most important field of application of the shallow-water equations. Computer models have been constructed for many areas in the world, from small to large scale. One possibility is the simulation of tidal flow on the entire globe, taking tide-generating forces by sun and moon into account. An example of the result for the semidiurnal component M_2 (period 12 h 25 min) is reproduced in fig. 1.2 (from Hendershott, 1981).

Many other questions could be answered using such numerical simulations, such as:
- what would happen to the tides in the North Sea if the Dover Strait were closed?
- how accurately should the sea bottom be represented?
- where should measuring systems be located to provide the maximum amount of information for the validation of models and for prediction purposes?
The general tendency is to apply such numerical models to smaller and smaller areas and take more and more detail into account (see following sections).

1.4. Tidal mixing

In very detailed (fine-grid) tidal models, it is possible to trace particle tracks. In areas of complicated geometry, this gives an interesting insight in the spreading of water particles and the effective mixing caused by strongly sheared flows. An example due to Ridderinkhof (1990) is shown in fig. 1.3. Some water particles are marked at a certain instant of time and their positions are traced during the computation of the tidal flow. The geometry is that of a tidal inlet in The Netherlands. The bottom topography in the shallow area just inside the inlet is quite complicated. This gives rise to a very pronounced spreading: marked particles initially close to one another move very far apart in a relatively short time. This process explains an important part of dispersion of particles even without turbulent mixing being involved. The efficiency of the process is related to the size of the cloud of particles compared with the size of eddies in the residual (tidally averaged) flow field.

1.5. Residual currents

Although tidal flow, being driven by solar and lunar influences, is periodic in principle, water particles do not describe exactly periodic orbits. Nonlinear effects can give rise to a net displacement of water particles after a tidal period. This can also be characterized as a rectified flow, or as the result of the radiation stress, exerted by a tidal wave (see also section 1.10). The net or residual flow may be small, but it is quite important for the transport rate and direction of dissolved or suspended substances (natural tracers, waste materials, sediment).

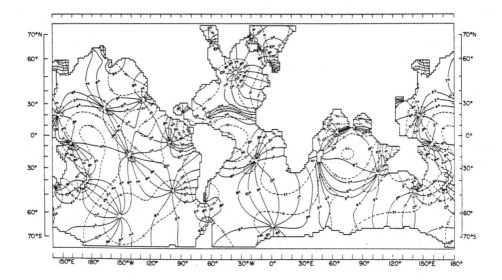

Fig. 1.2. Computed iso-phase lines (drawn) and iso-amplitude lines (dashed) for the semidiurnal tide in the world ocean (from Hendershott, 1981)

If the shallow-water equations are averaged over a tidal period, the nonlinear terms give rise to net momentum fluxes (radiation stresses) determined by the tidal flow component. If the fluxes are determined from the numerically computed tidal flows, the averaged equations could be solved for the residual currents, as originally proposed by Nihoul & Ronday (1975). These currents are the Eulerian residual currents, that is the flow velocity at a fixed location averaged over time. The problem is that this is not necessarily the velocity with which particles are transported. These are Lagrangean velocities, that could be determined by following particle tracks during a complete tidal cycle and connecting intial and final locations. There is a difference between Eulerian and Lagrangean residual velocities, as shown (a.o.) by Gerritsen (1985) from which fig. 1.4 is taken. Three particle patches are shown which are transported either by the Eulerian residual field or by the net particle displacement. The results show considerable differences. To complicate things, the Lagrangean displacement depends on the phase of the tide at which the particles are released.

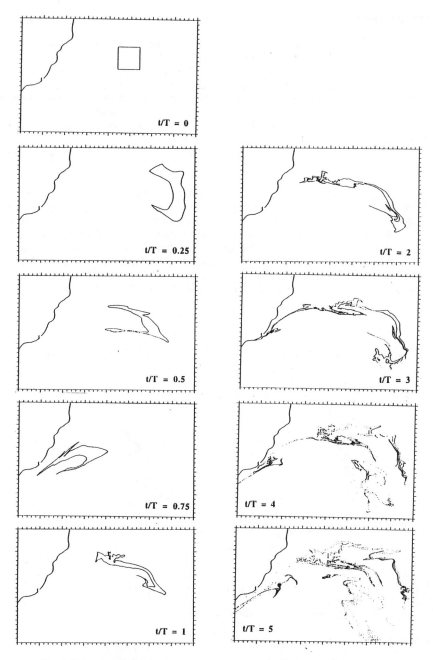

Fig. 1.3. Deformation of an initially square water column by tidal flow in a region with complicated bottom totpography. The shape of the water column is shown at indicated intervals (T is the tidal period) (from Ridderinkhof, 1990)

1.6. Storm surges

An important application of the SWE is the prediction of storm surges. Flows and water-level variations will be generated by atmospheric pressure differences and wind stresses on the water surface. The reliability of the predicted storm levels depends essentially on the quality of these meteorological input data. Using modern meteorological models, pressure distributions and wind fields can be generated with much detail even in regions where direct observations are scarce (i.e. at sea which is exactly where the data are most needed), and several days in advance. Moreover, usually the tide must also be taken into account because there is a nonlinear interaction between tides and storm surges: the two effects cannot just be superposed. The skill of storm surge models can be assessed by hindcasting former storms.

An example is given in fig. 1.5; it represents a hindcast by Dube et al (1985) for the November 1970 hurricane in the Bay of Bengal. The windfield was schematized from observed global data on the intensity and track of the storm. For an actual prediction, these would have to be predicted as well. The hydrodynamic model was based on the shallow-water equations, using curvilinear grids: the distance from the shore was normalized between 0 and 1. This also allows taking the deformation of the coastline by flooding into account (Johns et al 1982). The figure shows computed sea-level variations at a number of coastal stations and the contours of sea level at a particular instant of time in the model region.

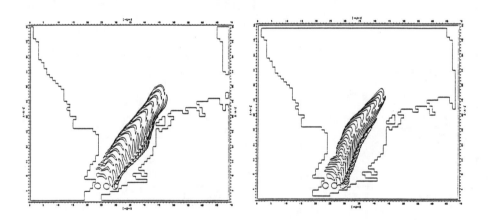

Fig. 1.4. Three initially circular patches are shown at intervals of 100 hrs.
(a) using a cyclic tidal velocity field;
(b) using the Eulerian residual velocity field (from Gerritsen, 1985)

1.7. River flows

Rivers with their flood plains are typical examples of shallow water. Flood waves in rivers are often very slowly varying (duration of several days) which causes them to behave differently from, e.g., tidal flows. The propagation speed of flood waves is small, of the same order as the flow velocity, and consequently considerably smaller than the speed expected for tidal waves. This can be explained by the fact that bottom friction is a dominant effect in this case. Fig. 1.6. (from Ogink et al, 1986) shows a small part of the River Waal in the Netherlands. The main river channel has extensive flood plains on both sides, surrounded by dikes. A numerical simulation of the flow may be used to study the dike levels required to withstand extreme floods, or the consequences of constructions of roads or bridges in the flood plain area.

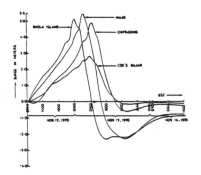 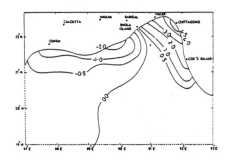

Fig. 1.5. Simulation of a hurricane near the Benghal coast.
(a) Time variation of predicted surface level at coastal stations;
(b) Contours (in m) of surface elevation on 13 November 0520 BST (from Dube et al, 1985)

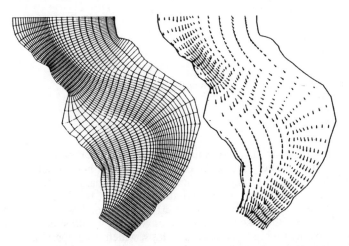

Fig. 1.6. Simulated flow in a river with flood plains. The numerical grid is shown, together with computed flow vectors (from Ogink et al, 1986)

1.8. Flows around structures

There has been a tendency to apply shallow-water models to flows at smaller and smaller scales: starting from tidal and atmospheric flows or hurricanes down to detailed flow patterns around hydraulic structures. An example of the latter was given by Stelling (1983). He studied the construction phase of a storm-surge barrier in one of the Dutch estuaries. The water depth varied from about zero on tidal flats to some 40 m in the deep channels. A rectangular grid with 90 m grid size was used, which stretches the applicability of the SWE to about its limits. An example of the tidal flow pattern in the flood phase is shown in fig. 1.7. It clearly shows flow contraction, lee areas and eddy formation. For more detailed information, 3-d models would have to be used.

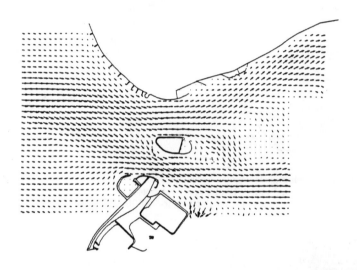

Fig. 1.7. Flow in an estuary during construction phase of storm-surge barrier (from Stelling, 1983)

1.9. Dambreak waves

With certain precautions, flows with discontinuities can also be computed numerically. Such flows do occur, though not very aften. An example is a tidal bore observed in some rivers. Another case is the wave resulting from the breaking of a dam. An example of the latter is given in fig. 1.8. It is related to a pumped-storage project in a lake or sea, where water is pumped up into a storage basin to a level of 20 m above sea level. The simulation illustrates what would happen if the dike around the storage basin would suddenly break. A series of pictures shows that a moving steep front develops, which is comparable to a shock wave in aerodynamics. It reflects at boundaries and gives significant water level rises there. For recent numerical simulations see, e.g. Alcrudo & Garcia-Navarro (1993).

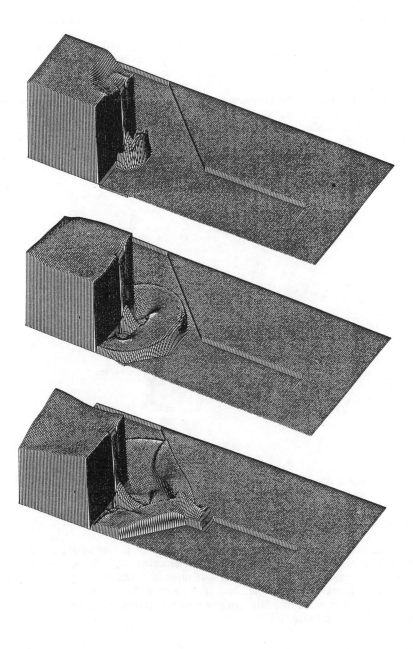

Fig. 1.8. Simulation of water level after breaking of a dam surrounding a reservoir (courtesy of Bosselaar, personal comm.)

1.10. Coastal flows

Currents along the coast are obviously influenced by tides and wind but they can also be driven by short waves. The latter, due to nonlinear effects, produce a net momentum flux or radiation stress, which acts as a driving force for the mean flow. The effect is particularly strong in the coastal wave-breaking region. If obstructions such as piers or breakwaters are present, they may deflect the coastal currents and cause strong seaward jets (rip currents). An example of this phenomenon, simulated by Wind & Vreugdenhil (1986) is given in fig. 1.9. A relatively simple model was used to compute the short-wave behaviour and the radiation stresses. These were input as external forces in a shallow-water model. In the breaker region, a strong longshore current is generated, which is turned into a seaward rip current at the downstream breakwater.

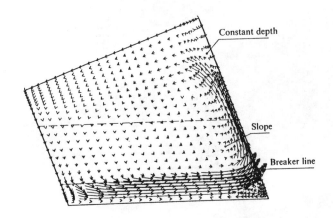

Fig. 1.9. Wave-driven flow along a coast with breakwaters (at left and right boundaries) (from Wind & Vreugdenhil, 1986)

1.11. Tsunamis

Tsunamis are ocean waves generated by undersea landslides or earthquakes. They may travel all the way across the ocean and grow when entering shallow border seas. The Japanese archipelago regularly experiences their sometimes devastating effects. Due to their large wave length, tsunamis belong to the class of shallow-water waves and can therefore be simulated using shallow-water models. An example by Shokin & Chubarow (1980) for the Japanese region is given in fig. 1.10.

1.12. Lake flows

If wind blows over a lake, this will generate flows in the lake. When the wind varies in time, oscillations of the lake level will result (Platzman, 1972). Even in steady state, quite complicated flow patterns can result due to differences in depth. The water level will then be tilted so as to give a pressure gradient in global equilibrium with the wind stress. This pressure gradient works more effectively in deeper than in shallower parts, contrary to the wind stress which is roughly the same over the lake surface. Consequently, in the shallower parts the pressure gradient will be insufficient to counteract the wind stress and the flow will be largely in the wind direction. The converse is valid in deeper areas where a return flow will develop. This looks like a 3-d effect but it can be simulated using the 2-d SWE. An example is given by Simons (1980) for the Baltic Sea, which is almost an enclosed lake. The region shown in the picture is embedded in a coarser model of the whole Baltic. The flow after a strong wind from the Northeast is illustrated. The effect of flows aligned with and opposed to the wind is clearly seen.

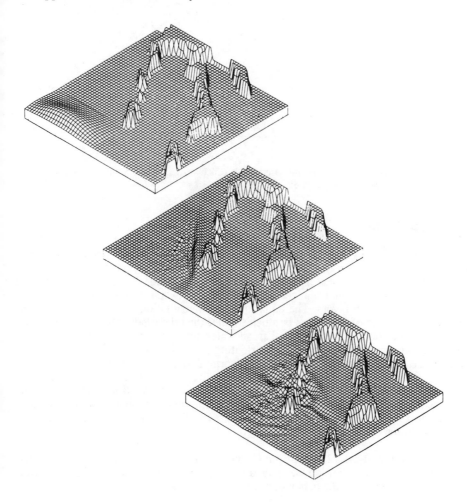

Fig. 1.10. Simulation of tsunami passing the Japanese islands
(from Shokin & Chubarow, 1980)

Fig. 1.11. Computed flow vectors and observations (thick arrows) for the Southwest Baltic after a strong northeast wind (from Simons, 1980)

1.13. Internal flows

If the fluid is stratified, e.g. due to differences in salinity, a layered flow results which is very similar to shallow-water flow. There are now two shallow layers, one on top of the other. In some cases, the water surface may be assumed to be fixed (rigid-lid approximation) and there will be changes in the interface between the layers only; however, there will be flows in both layers. The equations for this situation are very similar to the SWE, particularly if one of the layers is much thinner than the other. There may be a thin layer of salty water under a thick layer of fresh water acting as an "atmosphere", but the converse situation of a thin fresh-water layer floating on a thick layer of salty water acts in exactly the same way. The main difference with "normal" shallow-water flow is that gravity is reduced by a factor indicating the relative density difference between the layers, which may be as small as 0.001. This has a number of effects: the propagation speed of internal waves is much less than that of surface waves, but the amplitude of the waves may be much larger. Such flows are often indicated as reduced-gravity flows. In oceanography, this is a more or less common approximation.

An example is given in fig. 1.12 taken from Garvine (1987) showing the discharge of fresh water into the sea. Special attention was payed to the position of fronts. The Coriolis effect turned out to be important for the behaviour of the coastal flow.

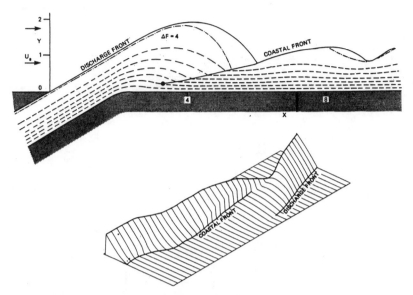

Fig. 1.12. Discharge of fresh water from a river into the sea.
The thickness of the upper (fresh-water) layer is shown as contour lines (top)
and in a 3-d view from below (bottom) (from Garvine, 1987)

1.14. Planetary flows

Some of the planets have an atmosphere which can be expected to work about the same way as the terrestrial one. As an example, a simulation of planetary flows on Jupiter is shown after Dowling & Ingersoll (1989). Their simulation was aimed at understanding the Great Red Spot on Jupiter as an atmospheric eddy. To this purpose, a reduced gravity model was set up similar to that in the preceding section (see also section 2.5.3). The atmosphere was supposed to consist of two layers. A deep, relatively heavy, lower layer was assumed to have a zonal (E-W) flow only and be in geostrophic equilibrium in the rotating system (the Coriolis parameter f is about twice its terrestrial value). On top of this, a relatively thin, lighter layer flows. The meridional (N-S) slope of the interface level provides the pressure gradient to support the equilibrium flow in the lower layer. This interface is hardly influenced by dynamic processes in the upper layer and therefore acts as a "bottom topography" for the latter.

A number of experiments were performed for a zonal channel. If a parallel flow was specified as an initial condition, with a profile derived from observations, it turned out to be unstable. Small eddies were formed which tended to coalesce, such that after about 5 to 10 years one large eddy developed that resembled the Great Red Spot. It then remained relatively stable for a long time. A sequence of pictures illustrating these events is shown in fig. 1.13. Lines of constant geopotential gh are given, in which reduced gravity g is poorly known as it involves the unknown density difference between the layers.

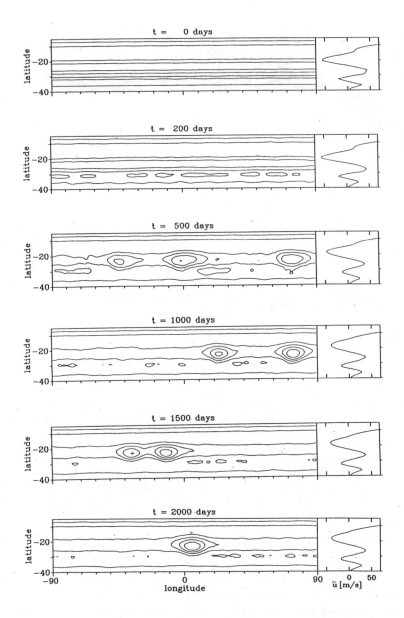

Fig. 1.13. Flow in a zonal channel on Jupiter, showing the instability of an initially specified zonally uniform flow, the development of eddies and their coalescence into one big eddy, identified with Jupiter's Great Red Spot (from Dowling & Ingersoll, 1989)

Chapter 2

Equations

Shallow-water flow is just one of the many special forms in which hydrodynamics presents itself. Actually, contrary to what the name suggests, the fluid does not have to be water. Certain aspects of flow in the atmosphere are described by the shallow-water equations (SWE) as well. The essential point is that the thickness a of the fluid layer is small compared with some typical horizontal length scale L. This is the only really common point in the various applications. There are several other parameters which may be either large or small, leading to variants of the SWE.

In order to see when and where a particular effect is important, it is necessary to derive the SWE with some care, so that you can see where approximations are needed and where they are not. It is supposed that you have a general background of hydrodynamics so that we can start at the general level of the Navier-Stokes equations.

2.1. Navier-Stokes and Reynolds equations

The Navier-Stokes equations describe conservation of mass and momentum. For the purpose of this book, we can limit the attention to incompressible fluids. This does *not* automatically mean that the fluid density is constant, but rather that it is independent of pressure p. The density may still vary due to other reasons, such as variations of temperature or salinity.

Conservation of momentum is expressed as

$$\frac{\partial}{\partial t}(\rho u) + \frac{\partial}{\partial x}(\rho u^2) + \frac{\partial}{\partial y}(\rho uv) + \frac{\partial}{\partial z}(\rho uw) - \rho fv + \frac{\partial p}{\partial x} - \frac{\partial \tau_{xx}}{\partial x} - \frac{\partial \tau_{xy}}{\partial y} - \frac{\partial \tau_{xz}}{\partial z} = 0$$

$$\frac{\partial}{\partial t}(\rho v) + \frac{\partial}{\partial x}(\rho uv) + \frac{\partial}{\partial y}(\rho v^2) + \frac{\partial}{\partial z}(\rho vw) + \rho fu + \frac{\partial p}{\partial y} - \frac{\partial \tau_{xy}}{\partial x} - \frac{\partial \tau_{yy}}{\partial y} - \frac{\partial \tau_{yz}}{\partial z} = 0 \tag{2.1}$$

$$\frac{\partial}{\partial t}(\rho w) + \frac{\partial}{\partial x}(\rho uw) + \frac{\partial}{\partial y}(\rho vw) + \frac{\partial}{\partial z}(\rho w^2) + \frac{\partial p}{\partial z} + \rho g - \frac{\partial \tau_{xz}}{\partial x} - \frac{\partial \tau_{yz}}{\partial y} - \frac{\partial \tau_{zz}}{\partial z} = 0$$

The coordinate system (fig. 2.1) is (x,y,z) with z positive upward and velocity components (u,v,w), t is time, p is pressure, ρ is density, g the acceleration due to gravity and $f = 2\Omega \sin \phi$ the Coriolis parameter, indicating the effect of the earth's rotation (Ω is the angular rate of revolution, ϕ the geographic latitude).

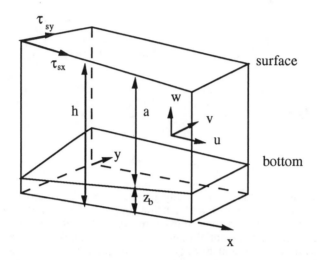

Fig. 2.1. Definition of coordinate system and boundaries

The viscous stresses τ_{ij} are expressed in terms of the fluid deformation rate as

$$\frac{\tau_{ij}}{\rho} = v \left(\frac{\partial u_i}{\partial x_j} + \frac{\partial u_j}{\partial x_i} \right) \tag{2.2}$$

where v is the kinematic viscosity (m²/s). To limit the amount of writing, the short-hand notation is used where x_j stands for (x,y,z) and u_j for (u,v,w) if $j = 1,2,3$.

Conservation of mass for a fluid element leads to the mass-conservation equation

$$\frac{\partial \rho}{\partial t} + \frac{\partial}{\partial x} (\rho u) + \frac{\partial}{\partial y} (\rho v) + \frac{\partial}{\partial z} (\rho w) = 0 \tag{2.3}$$

The density ρ of water depends in general on pressure p, temperature T and salinity S. Separate equations hold for conservation of heat and salinity (defined as the mass of dissolved salt per unit mass of water). The conservation of mass of dissolved salt gives

$$\frac{\partial}{\partial t} (\rho S) + \frac{\partial}{\partial x} (\rho u S) + \frac{\partial}{\partial y} (\rho v S) + \frac{\partial}{\partial z} (\rho w S) - \frac{\partial}{\partial x_j} (D \frac{\partial \rho S}{\partial x_j}) = 0 \tag{2.4}$$

where D stands for the diffusion coefficient for salt in water. If you combine this with the total mass equation (2.3) and assume the diffusion terms to be small (Batchelor, 1967, gives conditions for this to be true), you get the equation for salinity:

$$\frac{\partial S}{\partial t} + u \frac{\partial S}{\partial x} + v \frac{\partial S}{\partial y} + w \frac{\partial S}{\partial z} = 0$$

which says that the salinity of a certain package of water will remain constant if you move with it. A similar discussion applies to temperature. A fluid is said to be incompressible if its density does not depend on p. This does not mean that density is constant but that it depends only on T and S via the equation of state:

$$\rho = f(T,S) \tag{2.5}$$

Then the rate of change of density of a fluid element can be expressed as

$$\frac{d\rho}{dt} = \frac{\partial\rho}{\partial t} + u\frac{\partial\rho}{\partial x} + v\frac{\partial\rho}{\partial y} + w\frac{\partial\rho}{\partial z} = \frac{\partial\rho}{\partial T}\frac{dT}{dt} + \frac{\partial\rho}{\partial S}\frac{dS}{dt} = 0$$

again disregarding diffusion effects. Combining this with the conservation equation (2.3) for total mass, you finally get the mass balance equation (or continuity equation) for incompressible fluids:

$$\frac{\partial u}{\partial x} + \frac{\partial v}{\partial y} + \frac{\partial w}{\partial z} = 0 \tag{2.6}$$

So far, the discussion is independent of the exact form of the equation of state, which for water is a complicated relationship (cf., e.g. Gill, 1982), but the details are not needed here. Important is that for realistic temperature or salinity variations only small variations in density occur (usually less than a few percent). Such small variations have no important consequences in the inertia terms of (2.1) nor in the viscous terms, so that you can just take a constant density (that of fresh and/or cold water) there. The only point where density variations *are* important is in the gravity term in (2.1c), so there you must use the actual density. This approach of taking density variations into account only in the gravity term is called the *Boussinesq approximation* and it is commonly made in almost all kinds of geophysical flows.

In essentially all applications, you will find that the flow is turbulent, i.e. composed of stochastic motions or eddies on widely varying scales. Although the Navier-Stokes equations are generally believed to describe turbulence, that is not particularly useful as your interest will usually be in the large-scale features only. In order to isolate those, you may average the equations in some way (over an interval of time, over a domain of space or over an ensemble of possible stochastic realizations). This process is not free from fundamental difficulties for which you are referred to, e.g. Hinze (1975), Lumley & Panofsky (1964). Here we just suppose that each variable can be split into a slowly varying "mean" value and a "random" variation about it:

$$u = \bar{u} + u'$$

The important thing to note is that the mean of a product is not the product of the means, for example:

$$\overline{u\,v} = \bar{u}\,\bar{v} + \overline{u'v'}$$

If you substitute the splitting for all variables into the Navier-Stokes equations (2.1, 2.6) and take the average, you obtain what is called the Reynolds equations for the statistical average of a turbulent flow. These have the same form as the original Navier-Stokes equations; the difference is that additional stresses, called Reynolds stresses appear, representing the exchange of momentum between fluid elements by turbulent motion. They occur in exactly the same way as the viscous stresses, so we can combine them as

$$\frac{\tau_{ij}}{\rho} = v\left(\frac{\partial \overline{u}_i}{\partial x_j} + \frac{\partial \overline{u}_j}{\partial x_i}\right) - \overline{u'_i u'_j} \tag{2.7}$$

As they stand, the Reynolds stresses are unknown: they have to be expressed in terms of the mean motion to obtain a closed system of equations. This "closure problem" is one of the major tasks of turbulence research. For the present purpose it is sufficient to formulate the stresses in (2.7) in a way similar to (2.2) but with an effective turbulent or eddy viscosity v_t. The latter is found to be many orders of magnitude larger than the molecular one, and, worse, it is *not* a property of the fluid but rather of the flow. You will find some more discussion in section 2.7. In the following, we will use eqs. (2.1),(2.6), omitting the bars, but with the understanding that we are talking about the Reynolds-averaged quantities.

A similar question of averaging arises if the flow field includes short (wind-induced) waves. If you are interested in the large-scale flow features you would like to average over many short-wave periods, but again it turns out that the waves do have a net effect on the mean flow. The nonlinear terms give similar contributions as turbulence does in (2.7) but their signs are such that they are driving forces rather than "viscous" terms. These "radiation stresses" are discussed further in section 2.6.

2.2. Surface and bottom boundary conditions

In order to fix the solutions of the differential equations you need boundary conditions. For the derivation of the SWE, we will first of all use the conditions at the free water surface and at the solid bottom (fig. 2.1). Other boundary conditions are discussed in Chapter 5.

The surface and bottom conditions come in two kinds. The *kinematic* conditions say that water particles will not cross either boundary. For the solid bottom, this means that the normal velocity component must vanish:

$$u\frac{\partial z_b}{\partial x} + v\frac{\partial z_b}{\partial y} - w = 0 \quad \text{at } z = z_b \tag{2.8}$$

where z_b is the bottom level, measured from some horizontal reference level (fig. 2.1). At the free surface, things are a little more complicated as the surface may be moving by itself. Then the *relative* normal velocity must vanish:

$$\frac{\partial h}{\partial t} + u\frac{\partial h}{\partial x} + v\frac{\partial h}{\partial y} - w = 0 \quad \text{at } z = h \tag{2.9}$$

where h is the surface level measured from a horizontal reference level.

Secondly, we have *dynamic* boundary conditions which say something about the forces acting at the boundaries. At the bottom, you may assume that the viscous fluid "sticks" to the bottom, i.e.

$$u = v = 0 \qquad (2.10)$$

which is called the "no-slip" condition. Although this does not seem to specify any forces, it does so implicitly as you will see in section 2.7.

At the free surface, continuity of stresses is assumed, i.e. the stresses in the fluid just below the free surface are assumed to be the same as those in the air just above. This means that surface tension is not taken into account. For pressure, you find

$$p = p_a \qquad (2.11)$$

where p_a is the atmospheric pressure. The absolute pressure level is not important so it can be taken zero. Only the differences may be dynamically important, if you want to study the effect of atmospheric pressure variations on the water motion. On the sloping sea surface, a shear stress may act due to the wind. The shear stress (τ_{sx}, τ_{sy}) tangent to the water surface (fig. 2.1) is:

$$\tau_{sx} = -\tau_{xx}\frac{\partial h}{\partial x} - \tau_{xy}\frac{\partial h}{\partial y} + \tau_{xz} \quad \text{at} \quad z = h \qquad (2.12)$$

and similarly for the y direction. The wind stress vector on the left-hand side is supposed to be known as an external force (section 2.6). Then (2.12) with (2.2) (turbulent version) constitutes a boundary condition on the velocity field. At the bottom, a similar relation *defines* the bottom stress, which is not known and must be a result of the model (see section 2.7).

2.3. Scales

In order to specify what you mean by shallow water, you will have to consider typical scales. This is not so simple as it appears, as there are usually many different scales, either external ones, imposed by bottom topography or the variation of wind stress, or internal ones typical of wave lengths generated by the system itself.

Vertical length scales include:

(i) the water depth a (the symbol h is used for water surface *level*) ;

(ii) the thickness δ of boundary layers on the bottom or at the free surface; these are discussed to some extent in section 2.7. They are generated by viscosity and possibly influenced by the earth's rotation (in the latter case they are called Ekman layers). As they are determined by the flow, the boundary-layer thickness is an internal scale. In rivers, lakes and coastal seas, the boundary layers are usually thicker than the water depth, so actually the whole layer of water is within the boundary layer. In that case, the depth a is the relevant vertical scale. On the other hand, in very deep water (e.g. in the ocean), the boundary layers may be very thin in comparison with the water depth; then the water depth is still the relevant vertical scale for the overall motion.

(iii) the variation of bottom level: depth of channels, height of sand banks;

(iv) the variation of water level (i.e. wave amplitude); this is again an internal scale, if it is not directly imposed from the "outside world".

Horizontal length scales also come in various kinds:

(i) physical dimensions of the basin you are considering: width of a river or estuary, length of a harbour basin;

(ii) horizontal scales of bottom topography: width of channels, dimensions of sand banks;

(iii) the distance over which external forces vary significantly: variations of wind stress and atmospheric pressure in a storm;

(iv) the wave length, which is an internal scale determined by other factors such as the frequency of tidal forces and the dimensions of the basin (see ch. 4).

The basic assumption in shallow-water theory is that any vertical scale H is much smaller than any horizontal scale L. This puts the flows in the class of boundary-layer type flows. Just how small the ratio H/L should be is not easy to say. From linear wave theory, which does not involve the shallow-water assumption from the outset, it is found that a long shallow-water wave results if the ratio of water depth to wave length is less than about 0.05 (Le Méhauté, 1976). You can consider this to be about the upper limit of shallow-water theory.

For some purposes it is useful to introduce scales for the other variables as well, although this has nothing to do with shallow water. For velocity, there is usually only an internal scale U, not known beforehand. This will be a typical horizontal flow velocity. The vertical flow velocity will be smaller by a factor H/L as discussed in section 2.4.

For time, you may have an external scale, such as the tidal period, or an internal one, such as the wave period of a long shallow-water wave.

For viscosity, some estimate v of the turbulent eddy viscosity may be used as discussed in section 2.7. The molecular viscosity is usually negligible.

With these scales, three important dimensionless numbers may be formed, which determine the importance of several terms in the equations and, therefore, the type of flow:

Reynolds number $Re = UH/v$ ratio of inertia to viscous terms

Rossby number $Ro = U/fL$ ratio of inertia to Coriolis terms

Froude number $F = U/(ga)^{1/2}$ ratio of inertia to pressure-gradient terms

If turbulence is included in the definition of viscosity the Reynolds number is often in the range 100 to 1000. This is relatively large but insufficient to neglect turbulent friction (bottom friction) altogether. Depending on the length scale, the Rossby number is of the order 1 or above (rivers and coastal regions) to 10^{-3} (large-scale ocean currents). This means that the Coriolis acceleration is of minor importance in small-scale flows (it is often neglected) but dominating at large scales. The Froude number is usually small: it rarely exceeds 0.1 or 0.2. However, this is not small enough to neglect inertia terms.

2.4. Boundary-layer form

With the assumptions on the scales from the previous section, you can simplify the equations to a certain extent. First of all, look at the equation of continuity (2.6). If x,y have typical scales L, and the horizontal velocity components u,v are of the order U, each of the first two terms in (2.6) is of the order U/L. However, they will not usually cancel one another, which means that the third term must also be of that order; otherwise an equilibrium between the three is not possible. Assume the vertical length scale to be the water depth a, then the consequence is that the vertical velocity component is of the order $W = Ua/L$, i.e. it is smaller than the horizontal components by the same factor as the length scales.

The second step is made by considering the vertical momentum equation (2.1c). Using the length and velocity scales, you can very roughly estimate all terms except the pressure gradient. Due to the scaling of w, all advective terms turn out to be of comparable magnitude. You cannot neglect the vertical advection term as is sometimes done (in the following you will see that it is not necessary to do so either). The stress gradient terms with the viscosity formulation (2.2) involve second derivatives. Only the vertical derivatives are important in a boundary layer, as the second derivatives in horizontal direction are smaller by a factor $(a/L)^2$. Comparing everything with the gravitational term ρg, you will find the ratios:

local acceleration	advective terms	stress gradients
$\dfrac{Ua}{gTL}$	$\dfrac{U^2a}{gL^2}$	$\dfrac{vU}{ga^2}$

or

$\dfrac{Fa^2}{L^2}$	$\dfrac{F^2a^2}{L^2}$	$\dfrac{F^2}{Re}$

where it has been assumed (in anticipation of chapter 4) that $T = L/(ga)^{1/2}$. With the values of the dimensionless numbers from the previous section, you will conclude that all terms are small relative to the gravitational acceleration. Only the pressure gradient remains to balance it, and (2.1c) simplifies to the *hydrostatic pressure distribution*

$$\frac{\partial p}{\partial z} = -\rho g \tag{2.13}$$

This can be considered the central property in shallow-water theory (and more generally in boundary-layer theory). In (2.13), ρ is the actual density, as influenced by temperature and salinity. By integrating from the free surface using boundary condition (2.11) you get:

$$p = g \int_z^h \rho \, dz + p_a$$

If the density is not constant over depth, the pressure gradients will also depend on z. In this book, we focus on the depth-averaged SWE (see next section) in which such variations over depth cannot be accommodated. In that case, density must be assumed to be constant over the depth (though possibly varying in the horizontal plane), which gives as a special case

$$p = \rho g \, (h\text{-}z) + p_a \tag{2.14}$$

From (2.14), the pressure gradients in (2.1a,b) can be determined, e.g.

$$\frac{\partial p}{\partial x} = \rho g \frac{\partial h}{\partial x} + g(h\text{-}z)\frac{\partial \rho}{\partial x} + \frac{\partial p_a}{\partial x} \tag{2.15}$$

Collecting all results so far, the momentum equations become:

$$\frac{\partial u}{\partial t} + \frac{\partial}{\partial x}(u^2) + \frac{\partial}{\partial y}(uv) + \frac{\partial}{\partial z}(uw) - fv + g\frac{\partial h}{\partial x} + \frac{g}{\rho_0}(h\text{-}z)\frac{\partial \rho}{\partial x} + \frac{1}{\rho_0}\frac{\partial p_a}{\partial x} +$$

$$- \frac{1}{\rho_0}\{\frac{\partial \tau_{xx}}{\partial x} + \frac{\partial \tau_{xy}}{\partial y} + \frac{\partial \tau_{xz}}{\partial z}\} = 0$$

$$\tag{2.16}$$

$$\frac{\partial v}{\partial t} + \frac{\partial}{\partial x}(uv) + \frac{\partial}{\partial y}(v^2) + \frac{\partial}{\partial z}(vw) + fu + g\frac{\partial h}{\partial y} + \frac{g}{\rho_0}(h\text{-}z)\frac{\partial \rho}{\partial y} + \frac{1}{\rho_0}\frac{\partial p_a}{\partial y} +$$

$$- \frac{1}{\rho_0}\{\frac{\partial \tau_{yx}}{\partial x} + \frac{\partial \tau_{yy}}{\partial y} + \frac{\partial \tau_{yz}}{\partial z}\} = 0$$

This set, together with (2.6), could be called the "3-d shallow-water equations". They are not very much different from the standard boundary-layer equations. The x- and y derivatives of the stresses could be neglected, but it will turn out later on that this is not necessary for the depth-averaged SWE. Note that "prognostic" equations are available only for u and v: these variables can be "predicted" using eqs. (2.16). For w, only the "diagnostic" equation (2.6) can be used. The water level h will have to be determined from the boundary condition (2.9) or from the integrated equation of continuity (see next section).

2.5. Two-dimensional shallow-water equations

2.5.1. Integration over depth

The final step towards the 2-d SWE involves integration of the horizontal momentum equations (2.16) and the continuity equation (2.6) over depth $a = h - z_b$. To start with the latter, you get

$$\int_{z_b}^{h} (\frac{\partial u}{\partial x} + \frac{\partial v}{\partial y}) dz + [w]_{z_b}^{h} = \frac{\partial}{\partial x}(\overline{au}) + \frac{\partial}{\partial y}(\overline{av}) + [w]_{z_b}^{h} - \frac{\partial h}{\partial x} u_s - \frac{\partial h}{\partial y} v_s + \frac{\partial z_b}{\partial x} u_b - \frac{\partial z_b}{\partial y} v_b = 0$$

The overbar now indicates a depth average value. The surface (*s*) and bottom (*b*) correction terms come from the interchange of integration and differentiation with the boundary position depending on *x* and *y*. Using the boundary conditions (2.8) and (2.9) the bottom terms cancel and the surface terms yield the rate of change of the surface level:

$$\frac{\partial h}{\partial t} + \frac{\partial}{\partial x}(\overline{au}) + \frac{\partial}{\partial y}(\overline{av}) = 0 \tag{2.17}$$

This could have been derived directly by considering the mass balance for a vertical column of water.

A similar operation is applied to (2.16). The procedure for the advective terms is exactly the same as the one just shown and the surface and bottom terms now cancel fully due to the boundary conditions. However, you get nonlinear terms such as

$$\int_{z_b}^{h} uv \, dz = a \, \overline{u} \, \overline{v} + \int_{z_b}^{h} (u-\overline{u})(v-\overline{v}) dz$$

The second integral has some resemblance with the turbulent Reynolds stresses. However, it originates not from stochastic turbulent fluctuations, but from the deviations of the velocity field from its depth-averaged value. As it comes from the advection term, it can be called "differential advection term" and it describes a lateral momentum exchange due to differences in velocity over the depth of flow. This comes in excess of turbulent and molecular friction. Section 2.8 gives a further discussion.

The integration of the Coriolis and pressure terms is straightforward. The stress terms can be integrated as follows, e.g. for the *x*-momentum equation:

$$\int_{z_b}^{h} (\frac{\partial \tau_{xx}}{\partial x} + \frac{\partial \tau_{xy}}{\partial y} + \frac{\partial \tau_{xz}}{\partial z}) \, dz = \frac{\partial}{\partial x} \int_{z_b}^{h} \tau_{xx} \, dz + \frac{\partial}{\partial y} \int_{z_b}^{h} \tau_{xy} \, dz +$$

$$- [\tau_{xx}\frac{\partial h}{\partial x} + \tau_{xy}\frac{\partial h}{\partial y} - \tau_{xz}]_{z=h} + [\tau_{xx}\frac{\partial z_b}{\partial x} + \tau_{xy}\frac{\partial z_b}{\partial y} - \tau_{xz}]_{z=z_b}$$

The integrals are absorbed into the lateral stresses defined below; the boundary terms at surface and bottom yield the shear stress components in view of the boundary conditions (2.12). Here, you can see that it is not really necessary to neglect the τ_{xx}, τ_{xy} terms.

The result of the integration over the depth is:

$$\frac{\partial}{\partial t}(au) + \frac{\partial}{\partial x}(au^2) + \frac{\partial}{\partial y}(auv) - fav + ga\frac{\partial h}{\partial x} + \frac{ga^2\partial\rho}{2\rho_0\partial x} - \frac{1}{\rho_0}\tau_{bx} - \frac{\partial}{\partial x}(aT_{xx}) - \frac{\partial}{\partial y}(aT_{xy}) = F_x$$

$$\tag{2.18}$$

$$\frac{\partial}{\partial t}(av) + \frac{\partial}{\partial x}(auv) + \frac{\partial}{\partial y}(av^2) + fau + ga\frac{\partial h}{\partial y} + \frac{ga^2\partial\rho}{2\rho_0\partial y} - \frac{1}{\rho_0}\tau_{by} - \frac{\partial}{\partial x}(aT_{xy}) - \frac{\partial}{\partial y}(aT_{yy}) = F_y$$

Note that overbars indicating depth averages have been omitted. Together with (2.17), these equations form the shallow-water equations which are the subject of this book. The lateral stresses T_{ij} include viscous friction, turbulent friction and differential advection:

$$T_{ij} = \frac{1}{a}\int_{z_b}^h \left\{ v\left(\frac{\partial u_i}{\partial x_j} + \frac{\partial u_j}{\partial x_i}\right) - \overline{u'_i u'_j} + (u_i - \overline{u}_i)(u_j - \overline{u}_j) \right\} dz \tag{2.19}$$

In section 2.8 we will have a closer look at these lateral momentum-exchange processes. Driving forces $F_{x,y}$ include wind stress, atmospheric pressure gradient, and radiation stresses by (short) waves. Theyare discussed in section 2.6.

2.5.2. Splitting in vertical modes

The derivation so far is taylored to the situation where boundary layers are thick and cover the full water depth. This is the common situation in river and coastal engineering. In oceanography, the converse situation occurs where viscous influence is restricted to very thin layers near the surface and bottom. In that case, a form of the SWE can be derived by considering small disturbances from a stationary stratified inviscid fluid (cf. e.g. Gill, 1982). In this case, it is not necessary to assume that the density is uniform over the depth.

In the reference state, $u = v = w = 0$, $\rho = \rho_0(z)$ and $\partial p/\partial z = -\rho_0(z)g$. Assume small disturbances u' etc., forget about viscosity (which works only in the thin boundary layers) and linearize the equations:

$$\frac{\partial u'}{\partial t} + \frac{1}{\rho_0}\frac{\partial p'}{\partial x} - fv' = 0$$

$$\frac{\partial v'}{\partial t} + \frac{1}{\rho_0}\frac{\partial p'}{\partial y} + fu' = 0$$

$$\frac{\partial p'}{\partial z} = -\rho'g \tag{2.20}$$

$$\frac{\partial \rho'}{\partial t} + w'\frac{\partial \rho_0}{\partial z} = 0$$

$$\frac{\partial u'}{\partial x} + \frac{\partial v'}{\partial y} + \frac{\partial w'}{\partial z} = 0$$

You may now attempt a separation of variables. Assume that the vertical variation of u' is given by a, so far unknown, function $Z(z)$. Then, to satisfy the z-dependence in each of the equations, a consistent choice for the first four equations is

$$(u',v') = (u_1, v_1) \, Z(z)$$

$$p' = g \, h_1 \rho_0(z) \, Z(z)$$

$$\rho' = h_1 \frac{\partial}{\partial z} \{\rho_0(z) \, Z(z)\}$$

$$w' = gH \, w_1 \frac{\partial}{\partial z} \{\rho_0(z) \, Z(z)\} / \frac{\partial \rho_0}{\partial z}$$

in which the quantities subscripted by "1" are functions of x, y and t and H is a constant to be determined. If you put all this into (2.20 e), the separation is found to be possible only if $Z(z)$ satisfies

$$H \frac{\partial}{\partial z} \left(\frac{\partial}{\partial z} (\rho_0 Z) \, / \, \frac{\partial \rho_0}{\partial z} \right) - Z = 0 \qquad (2.21)$$

with appropriate boundary conditions at surface and bottom. This equation may be solved for Z only for special values of H; these are eigenvalues of the system. For each value, you obtain the following equations for the variables u_1, v_1, h_1:

$$\frac{\partial u_1}{\partial t} + g \frac{\partial h_1}{\partial x} - f v_1 = 0$$

$$\frac{\partial v_1}{\partial t} + g \frac{\partial h_1}{\partial y} + f u_1 = 0 \qquad (2.22)$$

$$\frac{\partial h_1}{\partial t} + H \left\{ \frac{\partial u_1}{\partial x} + \frac{\partial v_1}{\partial y} \right\} = 0$$

which are just special cases of the 2-d SWE with advective terms, surface and bottom stresses neglected and an effective or equivalent depth H. Note that the latter depends not only on the physical depth but also on the density distribution. Usually, one mode is found with H equalling the actual depth a, together with a number of "internal" modes with much smaller equivalent depths. In general, you will get all these modes superimposed.

2.5.3. Multilayer flow

Another case where a form of the SWE occurs is in two- or multilayer flow. Suppose that you have two layers of water with densities $\rho_{1,2}$. This may be a useful approximation in stratified estuaries or seas (e.g. Vreugdenhil, 1979). The integration process, described in section 2.5.1, can then be performed for each layer separately, taking into account exchange of mass and momentum across the interface. In the special case that there is no mass exchange, the following equations result (the top layer is indicated by "1", the bottom layer by "2"):

$$\frac{\partial a_1}{\partial t} + \frac{\partial}{\partial x}(a_1 u_1) + \frac{\partial}{\partial y}(a_1 v_1) = 0$$

$$\frac{\partial a_2}{\partial t} + \frac{\partial}{\partial x}(a_2 u_2) + \frac{\partial}{\partial y}(a_2 v_2) = 0$$

(2.23)

$$\frac{\partial}{\partial t}(a_1 u_1) + \frac{\partial}{\partial x}(a_1 u_1^2 + \tfrac{1}{2} g\, a_1^2) + \frac{\partial}{\partial y}(a_1 u_1 v_1) - f\, a_1 v_1 + g\, a_1 \frac{\partial a_2}{\partial x} + \tau_{ix} = 0$$

$$\frac{\partial}{\partial t}(a_2 u_2) + \frac{\partial}{\partial x}(a_2 u_2^2 + \tfrac{1}{2} g\, a_2^2) + \frac{\partial}{\partial y}(a_2 u_2 v_2) - f\, a_2 v_2 + g(1-\varepsilon)a_2 \frac{\partial a_1}{\partial x} - \tau_{ix} + \tau_{bx} = 0$$

with similar equations for y-momentum. Here, $\varepsilon = (\rho_2 - \rho_1)/\rho_1$ is the relative density difference (assumed constant) and τ_i the interfacial shear stress. The equations for the two layers are coupled, but in some special cases, the "normal" SWE are approximately recovered, for example if you add the corresponding equations for the two layers and neglect some small terms. Perhaps more interesting is the case where you have one thin layer (say, the lower one) with the other layer almost stagnant. Assuming u_1, v_1 to be small and applying $a_1 = h - a_2$ with the surface level h constant, the remaining equations for the lower layer are:

$$\frac{\partial a_2}{\partial t} + \frac{\partial}{\partial x}(a_2 u_2) + \frac{\partial}{\partial y}(a_2 v_2) = 0$$

(2.24)

$$\frac{\partial}{\partial t}(a_2 u_2) + \frac{\partial}{\partial x}(a_2 u_2^2 + \tfrac{1}{2}\varepsilon g\, a_2^2) + \frac{\partial}{\partial y}(a_2 u_2 v_2) - f\, a_2 v_2 - \tau_{ix} + \tau_{bx} = 0$$

and similarly for the y-direction. These equations are formally equivalent to the SWE, with the difference that the acceleration of gravity has been reduced by a factor ε. The effect is as if the lower layer acts as a single layer under an infinite atmosphere of density ρ_1. The model is known as the *reduced gravity model*. Almost the same result is obtained for a thin surface layer above a stagnant, thick, lower layer, a situation often occurring in oceanography. An example is given in section 1.13.

2.6. Driving forces

We have now established the main equations to be discussed in this book. It has become evident that there are a number of terms and coefficients in these equations that cannot be determined exactly. Estimates have to be made which may differ from one application to another. In the following sections 2.6 to 2.8, these terms are discussed in some more detail, with the dual purpose of
(i) estimating their importance, and
(ii) where possible giving useful semi-empirical formulations.

2.6.1. Atmospheric pressure gradient

In the driving forces on the right-hand side of eqs. (2.18), the atmospheric pressure gradient $\partial p_a/\partial x_i$ may be important for the simulation of storm surges (Heaps, 1967). The value of the gradient should be known from, e.g., a meteorological forecast.

2.6.2. Wind stress

The wind stress occurs as an important driving force in the SWE: the contributions to the right-hand side are $(\tau_{sx}, \tau_{sy})/\rho$. The magnitude and direction of the wind stress on the sea surface are determined by the flow in the atmosphere. Usually, the wind speed W is assumed to be known. An accepted semi-empirical formula for the magnitude of the wind stress is then given by Gill (1982):

$$\tau_s/\rho_{air} = c_f W^2 \tag{2.25}$$

and the direction is assumed to be the same as that of the wind. If the wind speed is measured at the 10 m level, the drag coefficient c_f is of the order 0.001. However, it is not quite constant: it still depends to some extent on the wind speed.

2.6.3. Density gradients

A third driving force, the density gradient, is shown on the left hand side of (2.18). It is useful to include this in the SWE only if the water mass is well mixed, such that the salinity or temperature, and thus the density is constant over the depth, but may still vary in horizontal direction. In cases where the 2-d SWE apply, its influence is usually small, the main effect being that a pressure (or water-level) gradient is set up to counteract the density gradient. In estuaries, this may produce a set-up of the water level in the order of several cm. In most cases, this is negligible compared with other effects and the density terms are therefore often neglected. If they are taken into account, the density distribution should either be known from observations, or an equation for it should be included, which is derived in the same way as the SWE. This is not further discussed in this book. If the density varies considerably over the depth, a 2-d model is no longer sufficient and you will have to apply the 3-d version of the SWE.

2.6.4. Radiation stress

In section 2.1 you have seen the splitting of velocity into mean and turbulent parts. This overlooks wave motions, which may produce driving forces for the mean flow. For example, in an oscillating boundary layer you will get a net flow in the direction of wave propagation ("boundary-layer streaming"). The same effect occurs in shallow water, which is a kind of boundary-layer flow as argued before. The analysis is described by Phillips (1977) for wind-generated waves after an original paper by Longuet-Higgins and Stewart (1964). A similar argument, applied to tidal waves, is reviewed by Robinson (1983).

A difficulty is that the water level varies with time, so that a direct average of the 3-d equations over time becomes troublesome: a particular control volume may be submerged only during part of the time. Therefore, the procedure is to average over depth first and then over time. Assume a three-part splitting:

$$u = u_m + u' = u_m + u'_w + u'_t$$

(with the subscripts indicating mean, waves and turbulence). The double averaging process then defines a mass flux by

$$M = U a + M_w \qquad (2.26)$$

$$\overline{U a} = \int_{z_b}^{h} u_m \, dz$$

$$M_w = \overline{\int_{z_b}^{h} u' \, dz} = \overline{\int_{z_b}^{h} u'_w \, dz}$$

The overbar indicates time averaging and it is assumed that the turbulent part does not have any net component. The mass transport due to waves M_w is not uniform over the depth of water but strongly concentrated near the surface (Battjes, 1988). You may now average the hydrodynamic equations in this way; for details see the references cited. You will get equations very much like (2.17, 18) but with a more complicated form of the lateral stresses T_{ij}. The assumption is usually made that the fluctuations due to waves and turbulence are not correlated, although this may be questionable in the coastal zone of wave breaking. The analysis by Phillips shows that the following terms, called radiation stresses, are added to (2.19) to account for the wave motion

$$T_{ij,w} = -\overline{\int_{z_b}^{h} (u'_{iw} u'_{jw} + \frac{p}{\rho} \, \delta_{ij}) \, dz} + \frac{1}{2} g a^2 \delta_{ij} \qquad (2.27)$$

(note the sign). This is still a very complicated expression, but it can be evaluated using linear wave theory. The result is

$$T_{ij,w} = -\frac{E c_g}{c} \frac{k_i k_j}{k^2} - \frac{E}{2} (\frac{2c_g}{c} - 1)\delta_{ij} \qquad (2.28)$$

with wave energy E, phase speed c, group velocity c_g defined as follows

$$E = \frac{1}{8} g H^2$$

$$c^2 = \frac{g}{k} \tanh ka$$

$$c_g = n\,c = (\tfrac{1}{2} + \frac{ka}{sinh\ 2ka})\,c$$

and with wave number k and wave height H. If the wave field is known, the radiation stresses act as driving forces for the mean flow, as indicated by the minus signs in (2.28). In this way, they have been applied to coastal wave driven currents by Wind and Vreugdenhil (1986) and others (cf. Battjes, 1988). All this assumes that the waves are not influenced by the mean flow, which may again be a questionable assumption.

2.6.5. Tidal stresses

It has been recognized that an effect similar to wave radiation occurs on a much larger scale in the generation of residual flows by oscillating tidal flows (Nihoul and Ronday, 1975, Zimmerman, 1978, Robinson, 1983). Various definitions of the residual flow have been proposed (e.g. Zimmerman, 1979, for a discussion see Robinson, 1983); the most convenient one appears to be that of eq. (2.26) which is called the Eulerian transport by Nihoul and Ronday. As tidal waves are just another class of waves, the analysis is the same as before. There is a modification only due to the bottom friction terms which may be important for tidal flows. If the bottom stress is modelled as a quadratic expression in terms of the depth-averaged velocity (see next section for a discussion), you get

$$\frac{\overline{\tau_{bx}}}{\rho} = \overline{c_f u \sqrt{u^2 + v^2}} \approx c_f u_m \overline{\sqrt{u_w^2 + v_w^2}} = c_f' u_m \tag{2.29}$$

assuming that the residual flow (subsript m) is small compared with the tidal component (subscript w). If the tidal flow is known, the result is a linear bottom stress for the residual flow with a modified coefficient depending on the tidal flow.

The idea was (Nihoul and Ronday, 1975) to compute the tidal flow first, using a numerical model. As the result usually is not a nice monochromatic wave, (2.28) does not apply and you may compute the radiation stresses (2.27) and bottom stress terms (2.29) directly by numerical averaging over time and then use these as known terms in the equations for the residual flow, which would then again be solved numerically. I do not enter into the discussion whether it is possible or useful to make this distinction; the point is that you should realize the existence of a net driving force from the tidal flow.

2.7. Bottom stress

The bottom stress is an unknown in the SWE, and it has to be expressed in terms of the other variables in order to "close" the system of equations. Essentially, specification of the bottom stress requires information on the 3-d flow structure. You can understand this by realizing that in the 3-d equations (2.16) the bottom stress is a *result* of the model when appropriate boundary conditions on the bottom are specified, such as no-slip conditions (2.10). In the SWE the bottom stress is specified in a parameterized form. Usually, it is assumed that the bottom stress depends quadratically on the depth-averaged velocity. This assumption comes from the analogy with equilibrium turbulent boundary layers, so it may be assumed to be correct of the shallow-water flow behaves as such.

There are, however, a few complicating factors, most importantly unsteadiness and rotation. In this section, you may get some insight in the validity of such a parameterization by studying a simplified 3-d situation. To this end, let us take a "quasi-laminar" model in which all non-linear effects have been neglected. The entire layer is assumed to be turbulent and we use a single constant eddy viscosity. The results of these simplifications are only semi-quantitative. From (2.16), with these approximations, you get

$$\frac{\partial u}{\partial t} - fv + g\frac{\partial h}{\partial x} = v\frac{\partial^2 u}{\partial z^2}$$

$$\frac{\partial v}{\partial t} + fu + g\frac{\partial h}{\partial y} = v\frac{\partial^2 v}{\partial z^2}$$

$$(2.30)$$

with no-slip boundary conditions at the bottom

$$u = v = 0 \qquad at \quad z = 0 \tag{2.31}$$

and a specified wind stress at the surface

$$v\frac{\partial}{\partial z}(u,v) = \frac{\tau_w}{\rho} \quad at \quad z = a \tag{2.32}$$

The *result* will then be the bottom stress

$$\frac{\tau_b}{\rho} = v\frac{\partial}{\partial z}(u,v) \quad at \quad z = 0 \tag{2.33}$$

This set of equations can be solved under rather general conditions, but for the present purpose, you may take some special situations.

2.7.1. Unsteady flow

In many cases, shallow-water flows are unsteady (tides, oscillations, storm surges). What is the behaviour of bottom stress then? Suppose that there is no wind stress, that Coriolis effects can be neglected and that the pressure gradient is given for a travelling wave. You may take the *x*-direction in the direction of the gradient, so that nothing happens in the *y*-direction.

$$g\frac{\partial h}{\partial x} = ikg\ H\ exp(ikx - i\omega t) \tag{2.34}$$

where *k* is the wave number, ω the frequency and *H* the amplitude. Then the velocity profile will also behave periodically like

$$u = U(z)\ exp(ikx - i\omega t)$$

with a velocity profile $U(z)$ to be determined. Inserting this into the equations (2.30) shows that $U(z)$ should satisfy

$$-i\,\omega\,U + ik\,g\,H = v\,U_{zz}$$

Using the boundary conditions (2.31, 2.32) the solution is found to be

$$U = \frac{gkH}{\omega}\left[1 - \frac{e^{\alpha(z-a)} + e^{-\alpha(z-a)}}{e^{\alpha a} + e^{-\alpha a}}\right]$$

where

$$\alpha = \sqrt{\frac{\omega}{2v}}\ (1-i)$$

The corresponding bottom stress is found from (2.33):

$$\frac{\tau_b}{\rho} = \frac{gkH\ v\alpha}{\omega}\ \frac{e^{\alpha a} - e^{-\alpha a}}{e^{\alpha a} + e^{-\alpha a}}\ exp(ikx - i\omega t) \tag{2.35}$$

The amplitude is a complex number different from U which means that the bottom stress is out of phase with the velocity. Usually, you would assume the bottom stress to be proportional to the depth-averaged velocity as in a steady state. If the flow would adjust to the pressure gradient in a quasi-steady way (without inertia), you would have

$$g\frac{\partial h}{\partial x} = v\frac{\partial^2 u}{\partial z^2}$$

and with (2.34) the solution for the U profile would be

$$U = \frac{igk\,H}{2v}\ z\,(z - 2a)$$

which is the well-known parabolic profile for laminar flow in a channel. The bottom stress would then be

$$\frac{\tau_b}{\rho} = -\,igk\,Ha\ exp(ikx - i\omega t) \tag{2.36}$$

which is now in phase with the velocity. The interesting thing is now to compare the dynamic and quasi-steady bottom stresses (2.35) and (2.36):

$$\frac{\tau_{b,\,dyn}}{\tau_{b,\,st}} = \frac{1}{\alpha a}\ \frac{e^{\alpha a} - e^{-\alpha a}}{e^{\alpha a} + e^{-\alpha a}} \tag{2.37}$$

This is shown in fig. 2.2 as a function of the parameter $\omega a^2/v$. The effect on the phase is more important than that on the amplitude. If the parameter is small, the ratio in (2.37) approaches unity, which means that the bottom stress responds to the oscillatory flow as though it were steady at each moment. You observe that this happens if $\omega a^2/v < 0.5$

approximately, or $a^2/v < 0.08\ T$ where T is the period of the flow. The quantity a^2/v can be interpreted as the time needed for a viscous flow to adjust its flow profile

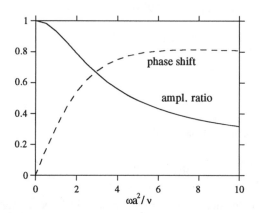

Fig. 2.2. Comparison of bottom stress in unsteady and quasi-steady flow. Drawn: amplitude; dashed: phase difference.

over a depth a. Therefore, you get a quasi-steady bottom stress if this adjustment time is less than roughly 10 % of the period of the oscillating flow.

Now for turbulent flow in shallow water, an estimate of the eddy viscosity is $v = 0.1$ m^2/s. For a depth $a = 10$ m the adjustment time will be of the order of half an hour, which is sufficiently small compared with the time scale for tidal flow of 12.5 h for the quasi-steady expression to apply. However, for smaller wave periods, the quasi-steady assumption may break down.

Zielke (1968) used a similar analysis in order to generate the formula for the wall stress in a pipe under transient flow conditions. However, it is not easy to generalize his approach to turbulent shallow-water flow.

2.7.2. Wind influence and rotation

In case of steady flow, there must be a driving force. Let us assume that this is a wind stress at the surface. We now include the effect of rotation. The two eqs. (2.30) can be combined into one by introducing the complex variables

$$Z = u + iv, \quad S = \frac{\partial h}{\partial x} + i\frac{\partial h}{\partial y}, \quad T = (\tau_{wx} + i\ \tau_{wy})/\rho$$

The equations then become

$$v\ Z_{zz} = ifZ + gS$$

$$(2.38)$$

$Z = 0$ at $z = 0$,
$v\,Z_z = T$ at $z = a$

The solution is

$$Z(z) = \frac{gS}{if}\left(-1 + \frac{e^{\alpha(z-a)} - e^{-\alpha(z-a)}}{e^{\alpha a} + e^{-\alpha a}}\right) + \frac{T}{\alpha v}\frac{e^{\alpha z} - e^{-\alpha z}}{e^{\alpha a} + e^{-\alpha a}}$$

where

$$\alpha = (1 + i)\sqrt{\frac{f}{2v}} = \frac{1+i}{a_E}$$

(this is a different parameter from the one in section 2.7.1). The quantity $a_E = \sqrt{2v/f}$ is called the Ekman depth. Its order of magnitude at middle latitude is 50 m, so it may be larger than the water depth in shallow areas.

The depth-averaged velocity can be determined by integrating (2.38a) over depth, tkaing into account the boundary condition (2.38c) and the definition of bottom stress (2.33)

$$\overline{Z} = -\frac{gS}{if} + \frac{T - T_b}{v\alpha^2 a}$$

The bottom stress follows from the velocity profile:

$$T_b = v[Z_z]_{z=0} = \alpha v\left\{\frac{gS}{if}(e^{-\alpha a} - e^{\alpha a}) + \frac{2T}{\alpha v}\right\}/(e^{\alpha a} + e^{-\alpha a})$$

You may now eliminate the pressure gradient S, which is not known a priori, in order to express the bottom stress in terms of the depth-averaged velocity. It is obvious that the wind stress will also enter into this relationship. The wind stress therefore not only has a direct effect as a driving term in the SWE, but it also enters indirectly in the resisting bottom stress. The result is (in vector form)

$$\frac{\tau_b}{\rho} = P\frac{v}{a}\overline{v} - Q\frac{\tau_w}{\rho} \tag{2.39}$$

with coefficients P and Q depending only on αa, that is, on $a^2 f/v$:

$$P = \frac{(\alpha a)^2 (e^{\alpha a} - e^{-\alpha a})}{\alpha a (e^{\alpha a} + e^{-\alpha a}) - (e^{\alpha a} - e^{-\alpha a})}$$

$$Q = \frac{e^{\alpha a} - e^{-\alpha a} - 2\alpha a}{\alpha a (e^{\alpha a} + e^{-\alpha a}) - (e^{\alpha a} - e^{-\alpha a})}$$

shown in fig.2.3. If $a^2 f/v$ is very small, you get a simple relation

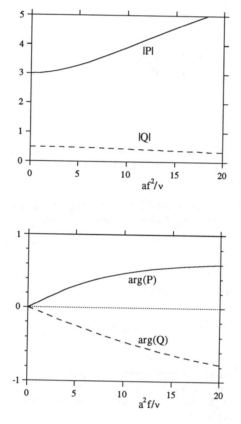

Fig. 2.3. Coefficients P and Q in (2.39). Top: amplitude; bottom: phase shift

$$\frac{\tau_b}{\rho} = 3\frac{\nu}{a}\overline{\mathbf{v}} - \frac{1}{2}\frac{\tau_w}{\rho} \tag{2.40}$$

which can more easily be derived by neglecting the Coriolis terms from the outset. If there is no wind stress, the wall stress is found to be proportional to the depth-averaged velocity, similar to an equilibrium laminar boundary layer. You see from fig. 2.3. that the absolute values of P and Q do not vary too much, but the phases do; this means that the Coriolis influence on the *direction* of the bottom stress is important. With the estimates for a ν given before and $f \sim 10^{-4}$, you find $a^2f/\nu \approx 0.2$, which means that the Coriolis influence is small; for greater depths and smaller viscosity, the influence may be appreciable.

2.7.3. Turbulent flow

The analysis in the previous section was done for "quasi-laminar" flow because it is hard to do for real turbulent flow. However, some of the results can be generalized. Particularly, the estimate of section 2.7.1 on the effect of unsteadiness can be expected to be valid in order of magnitude if a realistic value of the eddy viscosity is used.

For non-rotating two-dimensional flow, Reid (1957) derived an expression similar to (2.39) in which the linear term in the velocity is replaced by a quadratic one. The analysis can be generalized to 3-d (still without rotation, Vreugdenhil, 1977) but the expressions get very complicated. A very useful simple approximation was given by Groen and Groves (1962), again without Coriolis influence:

$$\frac{\tau_b}{\rho} = c_f \bar{v} |\bar{v}| - \gamma \frac{\tau_w}{\rho} \tag{2.41}$$

They recommend a value of $\gamma \approx 0.05$. The first term on the right-hand side is analogous to the wall stress for an equilibrium turbulent boundary layer and the coefficient c_f is a standard friction coefficient, depending on the wall roughness. Do not confuse it with a similar coefficient for the water surface in (2.25). Eq. (2.41) is the turbulent analogue of eq. (2.40) and can be assumed to be valid under the same conditions, i.e. if $a << a_F$. For the case without wind, you get the expression that is used very generally in the SWE. Written in its components:

$$\frac{\tau_{bx}}{\rho} = c_f u \sqrt{u^2 + v^2}$$
$$\tag{2.42}$$
$$\frac{\tau_{by}}{\rho} = c_f v \sqrt{u^2 + v^2}$$

This says that the bottom stress has the same direction as the depth-mean velocity and depends quadratically on its magnitude.

From (2.41) and (2.25) you may also estimate when the wind stress will have a significant influence on the bottom stress: this is the case if (with u as a typical velocity)

$$\frac{\gamma \tau_w}{c_{f,bottom}\rho_{water}u^2} = \frac{\gamma c_{f,surface}\rho_{air}W^2}{c_{f,bottom}\rho_{water}u^2} \approx 10^{-4} \left(\frac{W}{u}\right)^2 >> 1$$

The wind term is therefore important if the wind speed exceeds some 100 times the flow velocity, which will not often be the case.

You may conclude that there is a semi-empirical justification for (2.42) but only under certain conditions. Allowance for wind influence can be made by (2.41) but if the rotation effect becomes important, no simple adjustment is known and a 3-d model might be needed.

2.8. Lateral momentum exchange

2.8.1. Order of magnitude

In section 2.5.1. you have seen that there are three forms of lateral momentum exchange: viscous and turbulent stresses and differential advection. Viscous stresses are negligible in any realistic circumstances. The order of magnitude of the remaining contributions can be estimated (Kuipers and Vreugdenhil, 1973) by assuming a simple velocity profile. A logarithmic profile would probably be the obvious choice, but the formulae are a little simpler if you use a power-law form:

$$u = V\frac{n+1}{n} (z/a)^{1/n}, \quad v = 0 \tag{2.43}$$

where V is the depth-mean velocity and $n \approx 6$ is a parameter related to the bottom roughness. The corresponding eddy viscosity, which is not too different from the parabolic profile leading to a logarithmic velocity, is

$$v = \kappa^2 V a \left(1 - \frac{z}{a}\right) \left(\frac{z}{a}\right)^{1 - 1/n}$$

where κ is Von Karman's constant (~ 0.4).

To estimate the terms of (2.19) you might assume that the turbulent (Reynolds) stresses will all be of the same order of magnitude as the bottom stress, which for the power-law profile is

$$\tau_b/\rho = \frac{(n+1)^2}{n^4} \kappa^2 V^2 \tag{2.44}$$

The order of magnitude of the differential advection parts can then be computed as (take, e.g., T_{11}; the others will be of the same order of magnitude)

$$-\frac{\rho V^2}{a} \int_0^a \left\{1 - \frac{n+1}{n} (z/a)^n\right\}^2 dz = -\frac{\rho V^2}{n(n+2)} \tag{2.45}$$

The ratio of the turbulent contribution (2.44) to this is

$$\frac{(n+2)(n+1)^2}{n^3} \kappa^2$$

For $n = 6$ and $\kappa = 0.4$, this may amount to 30 %, which shows that both contributions may be important (the estimate given by Kuipers and Vreugdenhil, 1973 is in error).

2.8.2. Turbulent part

The turbulent contribution may be estimated by analogy with "normal" 3-d turbulent flows, i.e. by assuming an eddy-viscosity formulation in terms of the depth-mean velocity gradients. In section 3.1 it is shown that the SWE are analogour to the equations for *compressible flow*, so we use a form similar to the viscous stresses for compressible flow (Batchelor, 1967)

$$T_{ij,\,t} = v_h \left(\frac{\partial \overline{u}_i}{\partial x_j} + \frac{\partial \overline{u}_j}{\partial x_i} - \frac{1}{2} \delta_{ij} \frac{\partial \overline{u}_k}{\partial x_k} \right) \quad i,j = 1,2 \tag{2.46}$$

where the summation convention is used. Written in its components:

$$T_{11t} = v_h \left(\frac{\partial \overline{u}}{\partial x} - \frac{\partial \overline{v}}{\partial y} \right) \qquad T_{22t} = - v_h \left(\frac{\partial \overline{u}}{\partial x} - \frac{\partial \overline{v}}{\partial y} \right)$$

$$T_{12t} = T_{12t} = v_h \left(\frac{\partial \overline{u}}{\partial y} + \frac{\partial \overline{v}}{\partial x} \right)$$

The consequence is that $T_{11t} + T_{22t} = 0$. The assumption underneath is that there is no viscous resistance to compression and that the isotropic normal stress is represented in the pressure.

The difficulty has been shifted to the estimation of the effective horizontal eddy viscosity v_h. For a two-dimensional (depth-averaged) jet, Lean and Weare (1979) took the viscosity coefficient proportional to the center-line velocity and the width of the jet (see also Rodi, 1980). A similar assumption was made by Vreugdenhil and Wijbenga (1982) for the mixing layer between a river bed and its flood plains. Ogink (1985) reviewed some experiments for this case. Wind and Vreugdenhil (1986) used standard estimates for a wall jet in computing depth-averaged wave-driven flow near a groyne. Although some success has been obtained, you should be careful in using such estimates as the process of turbulence in a 2-d sheet of water is essentially different from the usual 3-d turbulence. In this respect, see also section 3.3.

A basically more correct approach was proposed by Rastogi and Rodi (1978, see also McGuirk and Rodi, 1978) who applied the ideas of the k - ε model for 3-d turbulence by stating

$$v_h = c_\mu \frac{k^2}{\varepsilon}$$

where k is the turbulent kinetic energy and ε the rate of energy dissipation. They averaged the equations for k and ε over depth in the same way as the momentum equations. Here, again, they had to introduce some assumptions on turbulence production by the bottom stress, but in any case this yields an approach which does not violate the 3-d character of turbulence.

2.8.3. Differential advection

The differential advection part of the lateral stresses (2.19) is even more difficult to estimate than the turbulent part. The deviation from the depth-mean velocity is not a random but an organized flow and even the sign of the integral in (2.19) is not clear. That means that the term can act either as a driving force or a resisting force. Actually, both may happen at the same time in different regions. Therefore, it is generally not possible to model this part in the same way as (2.46). See also section 3.3. where it is shown that the driving force for a body of water within a closed streamline should come from the lateral stresses in some way (in the absence of external forces).

Differential advection terms have been computed from a 3-d model by Versteegh (1990). Fig.2.4. shows the flow into and out of a "harbour" next to a river. The differential stress T_{sn} has been computed from the 3-d flow profiles, where s is the streamwise and n the normal direction. Also shown is the shear $\partial \overline{v}_s / \partial n$ in the depth-averaged flow. The differential stress and the velocity shear both have a maximum in the harbour entrance, but they even have different signs in the top region of the figure, where a spiral flow occurs. An effective viscosity as in (2.46) would be negative in such areas. Therefore, using a constant positive viscosity coefficient would not work for this case. Actually, this is one of the motivations for using 3-d models.

Kalkwijk and De Vriend (1980) analysed the differential advection effect for the special case of flow in a river bend. In this case, the longitudinal velocity may be assumed to be approximately logarithmic. It is then possible to estimate the radial velocity component and from the two, T_{sn} can be computed, with s the centerline and r the radial directions:

$$T_{sn} = - k_{sn} \frac{a}{R} u_s^2$$

where R is the radius of curvature and k_{sn} a numerical coefficient. T_{sn} is always negative, whereas the deformation rate

$$\frac{\partial u_s}{\partial n} + \frac{\partial u_n}{\partial s} \approx \frac{\partial u_s}{\partial n}$$

changes sign when you go from the inner to the outer bend, so that a negative eddy viscosity would result in the inner river bend.

The conclusion is that the differential advection part, if important, cannot be modelled by an effective viscosity. In any but special cases, a 3-d model will then be needed for a correct flow simulation.

2.9. Forms of the shallow-water equations

The "standard" shallow-water equations are obtained by disregarding some of the complications and adopting some parameterizations you have seen in the last few sections. They are the ones discussed further in this book. Disregarding lateral stresses and driving forces and assuming the simplest possible expression for the bottom stress, you get from (2.17) and (2.18):

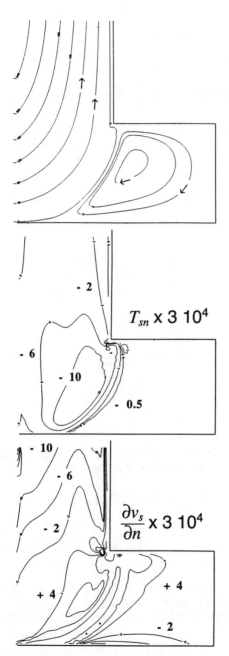

$T_{sn} \times 3 \ 10^4$

$\dfrac{\partial v_s}{\partial n} \times 3 \ 10^4$

Fig. 2.4. Streamlines, differential stress and velocity shear determined from a 3-d model (Versteegh, private comm.)

$$\frac{\partial h}{\partial t} + \frac{\partial}{\partial x}(au) + \frac{\partial}{\partial y}(av) = 0 \tag{2.48}$$

$$\frac{\partial}{\partial t}(au) + \frac{\partial}{\partial x}(au^2) + \frac{\partial}{\partial y}(auv) - fav + ga\frac{\partial h}{\partial x} + c_f u\sqrt{u^2 + v^2} = 0 \tag{2.49}$$

$$\frac{\partial}{\partial t}(av) + \frac{\partial}{\partial x}(auv) + \frac{\partial}{\partial y}(av^2) + fau + ga\frac{\partial h}{\partial y} + c_f v\sqrt{u^2 + v^2} = 0 \tag{2.50}$$

This is what is called the *conservative* form is at is written as far as possible in the form of a balance of fluxes of mass and momentum. The system can be written as a single vector equation

$$\frac{\partial \mathbf{v}}{\partial t} + \frac{\partial \mathbf{f}_1}{\partial x} + \frac{\partial \mathbf{f}_2}{\partial y} + \mathbf{s} = 0$$

with fluxes and sources defined as

$$\mathbf{v} = \begin{pmatrix} au \\ av \\ a \end{pmatrix} \quad \mathbf{f}_1 = \begin{pmatrix} au^2 + \frac{1}{2}g\,a^2 \\ auv \\ au \end{pmatrix} \quad \mathbf{f}_2 = \begin{pmatrix} auv \\ av^2 + \frac{1}{2}g\,a^2 \\ av \end{pmatrix}$$

$$\mathbf{s} = \begin{pmatrix} -fav + c_f u\,|\mathbf{v}| + ga\dfrac{\partial z_b}{\partial x} \\ fau + c_f v\,|\mathbf{v}| + ga\dfrac{\partial z_b}{\partial y} \\ 0 \end{pmatrix}$$

Often you will see another form which is obtained by performing the differentiations in (2.49,2. 50) and using (2.48) to cancel a number of terms:

$$\frac{\partial u}{\partial t} + u\frac{\partial u}{\partial x} + v\frac{\partial u}{\partial y} - fv + g\frac{\partial h}{\partial x} + c_f\frac{u}{a}\sqrt{u^2 + v^2} = 0 \tag{2.51}$$

$$\frac{\partial v}{\partial t} + u\frac{\partial v}{\partial x} + v\frac{\partial v}{\partial y} + fu + g\frac{\partial h}{\partial y} + c_f\frac{v}{a}\sqrt{u^2 + v^2} = 0 \tag{2.52}$$

which is the *differential or nonconservative* form of the SWE. Both forms are equivalent, but this is no longer so in numerical discretized form. There are a few applications where this may be important (shock waves, see section 3.4). Eqs. (2.48),(2.51),(2.52) can be written in vector form as well:

$$\frac{\partial \mathbf{v}}{\partial t} + A\frac{\partial \mathbf{v}}{\partial x} + B\frac{\partial \mathbf{v}}{\partial y} + C\,\mathbf{v} + \mathbf{s} = 0$$

with (note that the vector **v** of unknowns is different from the conservative case)

$$\mathbf{v} = \begin{pmatrix} u \\ v \\ a \end{pmatrix} \quad A = \begin{pmatrix} u & 0 & g \\ 0 & u & 0 \\ a & 0 & u \end{pmatrix} \quad B = \begin{pmatrix} v & 0 & 0 \\ 0 & v & g \\ 0 & a & v \end{pmatrix}$$

$$C = \begin{pmatrix} c_f |v|/a & -f & 0 \\ f & c_f |v|/a & 0 \\ 0 & 0 & 0 \end{pmatrix} \quad \mathbf{s} = \begin{pmatrix} g \dfrac{\partial z_b}{\partial x} \\ g \dfrac{\partial z_b}{\partial y} \\ 0 \end{pmatrix}$$

2.10. Curvilinear coordinates

Numerical methods based on curvilinear and possibly non-orthogonal grids are becoming more and more popular (see section 6.3). Therefore, it is useful to give a formulation of the equations in curvilinear coordinates. A formal coordinate transformation is introduced from the physical coordinates x,y into a new set of coordinates ξ, η :

$$\begin{aligned} \xi &= \xi(x,y) \\ \eta &= \eta(x,y) \end{aligned} \tag{2.53}$$

The differential equations are transformed using the chain rule. In conservation form, the equations will look like

$$\frac{\partial \mathbf{v}}{\partial t} + \frac{\partial \xi}{\partial x}\frac{\partial \mathbf{f}_1}{\partial \xi} + \frac{\partial \eta}{\partial x}\frac{\partial \mathbf{f}_1}{\partial \eta} + \frac{\partial \xi}{\partial y}\frac{\partial \mathbf{f}_2}{\partial \xi} + \frac{\partial \eta}{\partial y}\frac{\partial \mathbf{f}_2}{\partial \eta} + \mathbf{s} = 0 \tag{2.54}$$

In non-conservation form, you find

$$\frac{\partial \mathbf{v}}{\partial t} + A\left(\frac{\partial \xi}{\partial x}\frac{\partial \mathbf{v}}{\partial \xi} + \frac{\partial \eta}{\partial x}\frac{\partial \mathbf{v}}{\partial \eta}\right) + B\left(\frac{\partial \xi}{\partial y}\frac{\partial \mathbf{v}}{\partial \xi} + \frac{\partial \eta}{\partial y}\frac{\partial \mathbf{v}}{\partial \eta}\right) + C\mathbf{v} + \mathbf{s} = 0$$

or

$$\frac{\partial \mathbf{v}}{\partial t} + \left(A\frac{\partial \xi}{\partial x} + B\frac{\partial \xi}{\partial y}\right)\frac{\partial \mathbf{v}}{\partial \xi} + \left(A\frac{\partial \eta}{\partial x} + B\frac{\partial \eta}{\partial y}\right)\frac{\partial \mathbf{v}}{\partial \eta} + C\mathbf{v} + \mathbf{s} = 0 \tag{2.55}$$

which has the same form as the original untransformed equations but with more complicated coefficients.

So far, the transformation has been assumed independent ·of time, but you can use the same technique if the new coordinates move with respect to the old ones. This can be the case if you have moving boundaries (e.g. due to flooding and drying of tidal flats) which you want to have at a fixed position in the new coordinates. You will just get some more transformation terms:

$$\frac{\partial \mathbf{v}}{\partial t} + \left(\frac{\partial \xi}{\partial t} + A\frac{\partial \xi}{\partial x} + B\frac{\partial \xi}{\partial y}\right)\frac{\partial \mathbf{v}}{\partial \xi} + \left(\frac{\partial \eta}{\partial t} + A\frac{\partial \eta}{\partial x} + B\frac{\partial \eta}{\partial y}\right)\frac{\partial \mathbf{v}}{\partial \eta} + C\mathbf{v} + \mathbf{s} = 0 \quad (2.56)$$

In the following, only the case of a time-independent transformation is discussed.

The transformed equations (2.54)...(2.56) still contain the original Cartesian velocity components. It is possible to solve them that way but, alternatively, you may transform them to velocity components related to the curvilinear coordinate system. In order to avoid a messy notation, it is advisable to use standard tensor analysis, which you can find in any good book; here we follow Aris (1962). The Cartesian coordinates (x,y) are indicated by x^j ($i = 1,2$), the curvilinear ones (ξ,η) by ξ^j ($j = 1,2$). You can define two sets of base vectors in the curvilinear system (see fig.2.5). The covariant base vectors (which do not have unit length) are tangent to coordinate lines:

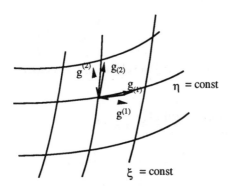

Fig. 2.5. Definition of covariant and contravariant base vectors

$$\mathbf{g}_{(i)} = \frac{\partial \mathbf{x}}{\partial \xi^i} \quad or \quad \{g_{(i)}\}_j = \frac{\partial x_j}{\partial \xi^i}$$

The reciprocal or contravariant base vectors $\mathbf{g}^{(i)}$ are normal vectors to grid lines (indicated by a *superscript*) . The definition is in components

$$g_j^{(i)} = \frac{\partial \xi^i}{\partial x^j}$$

They form a bi-orthogonal system with the covariant ones:

$$\mathbf{g}_{(i)} \cdot \mathbf{g}^{(j)} = \delta_j^i$$

(the Kronecker symbol $\delta_j^i = 1$ if $i = j$, otherwise 0). That this is so, you can see by using (2.53). Differentiating it with respect to ξ and η, it is straightforward to find the relations

$$
\begin{aligned}
J\,\xi_x &= y_\eta & J\,\eta_x &= -\,y_\xi \\
J\,\xi_y &= -\,x_\eta & J\,\eta_y &= x_\xi
\end{aligned}
\tag{2.57}
$$

where the Jacobian J is defined as

$$J = x_\xi y_\eta - x_\eta y_\xi$$

Using this, you can check , e.g.

$$\mathbf{g}_{(1)} \cdot \mathbf{g}^{(2)} = \frac{\partial x^1}{\partial \xi^1}\frac{\partial \xi^2}{\partial x^1} + \frac{\partial x^2}{\partial \xi^1}\frac{\partial \xi^2}{\partial x^2} = x_\xi\,\eta_x + y_\xi\,\eta_y = J\,(\eta_x\,\eta_y - \eta_y\,\eta_x) = 0$$

Metric quantities are defined as

$$
\begin{aligned}
g_{ij} &= \mathbf{g}_{(i)} \cdot \mathbf{g}_{(j)} \\
g_{11} &= x_\xi^2 + y_\xi^2 & g_{22} &= x_\eta^2 + y_\eta^2 \\
g_{21} &= g_{12} = x_\xi y_\xi + x_\eta y_\eta
\end{aligned}
\tag{2.58}
$$

Similarly,

$$g^{ij} = \mathbf{g}^{(i)} \cdot \mathbf{g}^{(j)}$$

The determinant of the matrix g_{ij} is indicated by $g*$ (in tensor analysis the usual notation is g; the asterisk avoids confusion with gravity).

Denote the velocity vector (u, v) in the Cartesian system as u^i. In the transformed system you have the choice between two sets of velocity components. The covariant ones are those along the ξ_j grid lines; the contravariant ones are normal to the grid lines. In the general non-orthogonal case, the two are different. In chapter 6, you will see that the normal fluxes on the cell faces are needed, so it is convenient to use the contravariant components indicated by v^j. The transformation relations between them and the Cartesian components are

$$u^i = \frac{\partial x^i}{\partial \xi^j}\,v^j \quad \text{and} \quad v^j = \frac{\partial \xi^j}{\partial x^k}\,u^k \tag{2.59}$$

(the summation convention is used unless otherwise specified). As the base vectors are not unit vectors (and possibly not even dimensionless), the components v^j are not the physical velocity components $v(j)$, that is, the actual velocity components in the coordinate directions. To get those, you have to scale them

$$v(j) = \sqrt{g_{jj}}\ v^j \qquad (2.60)$$

(no summation on j).

In the curvilinear system, partial derivatives of scalars keep their usual significance. This is not so for vectors where the derivatives get a more complicated form as the base vectors change when you move across the field. For any contravariant vector field a^j, the covariant derivative in the k-direction is defined as

$$a^j_k = \frac{\partial a^j}{\partial \xi^k} + \left\{ \begin{array}{c} j \\ i\ \ k \end{array} \right\} a^i \qquad (2.61)$$

where the Christoffel symbols of the second kind are defined as

$$\left\{ \begin{array}{c} j \\ i\ \ k \end{array} \right\} = \frac{\partial \xi^j}{\partial x^p} \frac{\partial^2 x^p}{\partial \xi^i \partial \xi^k} = \tfrac{1}{2} g^{jp} \left(\frac{\partial g_{pi}}{\partial \xi^k} + \frac{\partial g_{pk}}{\partial \xi^i} - \frac{\partial g_{ik}}{\partial \xi^p} \right)$$

Note that these involve second derivatives of the transformation functions (2.53).

The hydrodynamic equations are derived by Aris (1962) directly from mass and momentum balances in the transformed system. The equation for mass continuity becomes (in the present notation)

$$\frac{\partial h}{\partial t} + (av^i)_{,i} = 0 \qquad (2.62)$$

This can be somewhat simplified by taking account of the relation

$$\left\{ \begin{array}{c} i \\ j\ \ i \end{array} \right\} = \frac{1}{\sqrt{g^*}} \frac{\partial \sqrt{g^*}}{\partial \xi^j}$$

so

$$\frac{\partial h}{\partial t} + (av^i)_{,i} = \frac{\partial h}{\partial t} + \frac{\partial}{\partial \xi^i}(av^i) + \frac{1}{\sqrt{g^*}} \frac{\partial \sqrt{g^*}}{\partial \xi^i} av^i = \frac{\partial h}{\partial t} + \frac{1}{\sqrt{g^*}} \frac{\partial}{\partial \xi^i}(\sqrt{g^*}\ av^i) = 0 \qquad (2.63)$$

The momentum equations are given by Aris in the nonconservative form:

$$\frac{\partial v^i}{\partial t} + v^j v^i_j + g\, g^{ij} \frac{\partial h}{\partial \xi^j} = f^i \qquad (2.64)$$

where the right-hand member f^i contains Coriolis and bottom friction terms:

$$f^i = -\frac{c_f}{a} v^i |v| - f \varepsilon_{ik} g_{kj} v^j \tag{2.65}$$

You can check this by noting that the friction term is supposed to work in the direction of the velocity vector and the Coriolis acceleration normal to it. Here,

$$\varepsilon_{ik} = \begin{pmatrix} 0 & -1 \\ 1 & 0 \end{pmatrix}$$

Complications compared with the equations in Cartesian coordinates are threefold:
(i) Christoffel symbols occur in the advective terms,
(ii) the pressure term in the i -th equation is generally not just the gradient in that coordinate direction,
(iii) the Coriolis term is more complicated.

The conservative form is easily recovered by multiplying (2.62) by v^i, (2.64) by depth a and adding. You should realize that the normal product rule

$$(av^iv^j)_k = av^jv^i_k + v^i(av^j)_k$$

applies to covariant derivatives if you use the proper form of a derivative of a second-order tensor:

$$(av^iv^j)_k = \frac{\partial}{\partial \xi^k} (av^iv^j) + \left\{ \begin{matrix} i \\ p \ \ k \end{matrix} \right\} av^pv^j + \left\{ \begin{matrix} j \\ p \ \ k \end{matrix} \right\} av^iv^p \tag{2.66}$$

Then, you find the momentum equations:

$$\frac{\partial}{\partial t} (av^i) + (av^iv^j)_j + ag \, g^{ij} \frac{\partial h}{\partial \xi^j} = af^i \tag{2.67}$$

which you may transform into physical components using (2.60) if you wish.

A special case is obtained if the coordinates are orthogonal. This is expressed by the condition the two covariant base vectors are orthogonal:

$$g_{12} = x_\xi x_\eta + y_\varsigma y_\eta = 0 \tag{2.68}$$

It is then customary to write

$$g_{ii} = (h_i)^2$$

(no summation). The mass-balance equation (2.63) keeps the same form as above, where now

$$g^* = \det(g_{ij}) = (h_1 h_2)^2$$

In the momentum equations, the Coriolis and pressure terms are simplified as $g_{12} = 0$, so

$$f^1 = -\frac{c_f}{a} v^1 |v| - f h_2^2 v^2$$
$$f^2 = -\frac{c_f}{a} v^2 |v| + f h_1^2 v^1$$

$$(2.69)$$

The Christoffel symbols simplify in the orthogonal case (Aris 1962):

$$\left\{ \begin{array}{cc} 1 \\ 1\ 1 \end{array} \right\} = \frac{1}{h_1} \frac{\partial h_1}{\partial \xi^1} \qquad\qquad \left\{ \begin{array}{cc} 1 \\ 2\ 1 \end{array} \right\} = \left\{ \begin{array}{cc} 1 \\ 1\ 2 \end{array} \right\} = \frac{1}{h_1} \frac{\partial h_1}{\partial \xi^2}$$

$$\left\{ \begin{array}{cc} 2 \\ 1\ 1 \end{array} \right\} = \frac{-h_1}{h_2^2} \frac{\partial h_1}{\partial \xi^2} \qquad\qquad \left\{ \begin{array}{cc} 1 \\ 2\ 2 \end{array} \right\} = \frac{-h_2}{h_1^2} \frac{\partial h_2}{\partial \xi^1}$$

$$\left\{ \begin{array}{cc} 2 \\ 1\ 2 \end{array} \right\} = \left\{ \begin{array}{cc} 2 \\ 2\ 1 \end{array} \right\} = \frac{1}{h_2} \frac{\partial h_2}{\partial \xi^1} \qquad\qquad \left\{ \begin{array}{cc} 2 \\ 2\ 2 \end{array} \right\} = \frac{1}{h_2} \frac{\partial h_2}{\partial \xi^2}$$

If you introduce the physical components (2.60) of the velocity vector, the continuity equation (2.63) becomes

$$\sqrt{g^*} \frac{\partial h}{\partial t} + \frac{\partial}{\partial \xi^1} \{h_2\, av(1)\} + \frac{\partial}{\partial \xi^2} \{h_1\, av(2)\} = 0 \qquad\qquad (2.70)$$

Similarly, from the momentum equations (2.64) you find

$$\frac{\partial v(1)}{\partial t} + \frac{v(1)}{h_1} \frac{\partial v(1)}{\partial \xi^1} + \frac{v(2)}{h_2} \frac{\partial v(1)}{\partial \xi^2} + \frac{v(1)v(2)}{h_1 h_2} \frac{\partial h_1}{\partial \xi^2} - \frac{v^2(2)}{h_1 h_2} \frac{\partial h_2}{\partial \xi^1} + g\, h_1 \frac{\partial h}{\partial \xi^1} = f^1$$

$$(2.71)$$

and similarly for the second equation. The terms involving derivatives of h_i can be interpreted as acceleration terms due to grid curvature.

Chapter 3

Some properties

3.1. Correspondence with compressible flow

The SWE have the same appearance as the Euler equations for compressible flow in aerodynamics. It is important to be aware of this similarity, because it opens the vast body of experience with the numerical solution of the Euler equations for application to the SWE.

The comparison to compressible flow may seem strange to you as the SWE were derived for incompressible flow. However, an apparent compressibility is caused by the presence of the free surface. To show the similarity, you may forget for a moment about external forces, Coriolis acceleration and bottom friction and assume a horizontal bottom. Then the depth a can be used as a variable instead of surface level h. It is now possible to *define* a density ρ by

$$\frac{\rho}{\rho_0} = \frac{a}{a_0}$$

with more or less arbitrary values a_0 and ρ_0; they are only there to ensure correct dimensions. Similarly, a pressure p can be *defined* by

$$p = p_0 + \frac{\rho_0 g}{2a_0} a^2$$

where p_0 is a constant reference pressure. Together with the definition of density, this becomes

$$p = p_0 + \frac{a_0 g}{2\rho_0} \rho^2$$

which formally is an equation of state, though not a very realistic one: real gases have a power of 5/3 on the density (for monatomic gases) or less, as opposed to an exponent of 2 in this case. Using these definitions, the equation of continuity (2.48) becomes

$$\frac{\partial \rho}{\partial t} + \frac{\partial}{\partial x}(\rho u) + \frac{\partial}{\partial y}(\rho v) = 0 \tag{3.1}$$

The pressure gradient in the momentum equation is transformed as

$$g \frac{\partial a}{\partial x} = \frac{g}{2a} \frac{\partial}{\partial x} (a^2) = \frac{1}{\rho} \frac{\partial p}{\partial x}$$

and consequently the x-momentum equation becomes

$$\frac{\partial u}{\partial t} + u \frac{\partial u}{\partial x} + v \frac{\partial u}{\partial y} + \frac{1}{\rho} \frac{\partial p}{\partial x} = 0 \tag{3.2}$$

(similarly for the y direction). Eqs. (3.1.) and (3.2) are exactly in the form of the Euler equations for compressible flow. There is one more difference except the equation of state: there is no separate energy equation available. The physical interpretation of the similarity with compressible flow is that convergence or divergence of the flow is possible and will be balanced either by density changes (in compressible flow) or by depth changes (in shallow-water free-surface flow).

The characteristic speed of disturbances in compressible flow (sound speed) is

$$c = \sqrt{\frac{dp}{d\rho}}$$

With the above equation of state, this becomes $c = \sqrt{ga}$ which agrees with the characteristic speed of long water waves to be discussed in chapter 4. Correspondingly, the Mach number $M = u/c$ is translated into the Froude number $F = u / \sqrt{ga}$. For many aerodynamic applications, the Mach number is not small and there are supersonic (supercritical) regions ($M > 1$) and shock waves. For spacecraft applications, the Mach number may even be much larger than 1. On the other hand, for most shallow-water flows are subcritical and F rarely exceeds 0.1 or 0.2. Only in special cases (e.g. flow over dams) can supercritical flow occur.

3.2. Correspondence with incompressible viscous flow

There is another similarity, now with the Navier-Stokes equations for incompressible flow in two dimensions, which applies if there are no important free-surface fluctuations. Chapter 4 indicates when this is the case. As in the previous section, it is useful to be aware of this similarity so that you can use the experience obtained in numerically solving the Navier-Stokes equations. Again assume for a while that there are no external forces, no Coriolis influence and that the bottom is horizontal (neither of these assumptions is essential and the corresponding terms can be added again).

As an additional and more important assumption, you have to make the *rigid-lid approximation*. This says that the free surface, if it varies only slightly, may be replaced by fixed, rigid and frictionless upper boundary or lid. You may no longer assume zero (or atmospheric) pressure on that artificial surface because a nonzero pressure p_s (positive or negative) will be exerted on such a fixed boundary. Actually, this pressure, if translated into a water column, indicates the level h the free surface would have attained if it were really free:

$$p_s = \rho g (h - h_s)$$

where h_s is the artificially fixed level. The latter need not even be horizontal or stationary, but we assume it is. The hydrostatic pressure distribution (2.14) is then replaced by

$$p = p_s + \rho g \, (h_s - z)$$

and consequently, independent of z

$$\frac{\partial p}{\partial x} = \frac{\partial p_s}{\partial x}$$

From eq.(2.48) you find now

$$\frac{\partial u}{\partial x} + \frac{\partial v}{\partial y} = 0 \tag{3.3}$$

which is the equation of continuity for 2-d incompressible flow. The momentum equation (2.51) becomes:

$$\frac{\partial u}{\partial t} + u\frac{\partial u}{\partial x} + v\frac{\partial u}{\partial y} + \frac{1}{\rho}\frac{\partial p_s}{\partial x} - v_h\left(\frac{\partial^2 u}{\partial x^2} + \frac{\partial^2 u}{\partial y^2}\right) = 0 \tag{3.4}$$

and similarly in the y direction. It has been assumed that a lateral stress according to eq. (2.46) can be used and that the viscosity coefficient is constant. Eqs. (3.3) and (3.4) are the Navier-Stokes equations for 2-d incompressible flow. Note that the surface pressure appears as an unknown that is evaluated together with the solution for u, v.

The analogy between Navier-Stokes and the SWE, again, has its limitations. It can be exploited only if the surface can be considered more or less fixed. This is first of all the case in steady flow (if spatial water level variations are small compared with depth). Secondly, in Chapter 4 you will see that there are certain wave types which have essentially a fixed surface. Thirdly, even if there are significant waves (say, tidal waves) but you are studying a small region compared with the wave length, you could consider the surface to be rigid but moving up and down in a prescribed way. A "source" term corresponding to this "pump" should then be included in (3.3). Examples can be found, e.g. in Officier et al (1986).

3.3. Conservation laws

The equations (2.48)...(2.50) are in the form of local conservation laws, indicating the rate of change of mass or momentum in case of a local unbalance of fluxes. You will get global conservation laws by integration of the local ones over a closed region.

3.3.1. Mass

By integrating (2.48) over an area A, enclosed by a boundary S, and applying Gauss' theorem, you obtain

$$\frac{\partial}{\partial t}\iint_A a \, dx \, dy + \int_S a \, v.n \, ds = 0 \qquad (3.5)$$

which says that the rate of change of total mass equals the net inflow rate across the boundary (**n** is the outward normal). If there is no inflow, the total mass is constant.

3.3.2. Momentum

A similar global balance can be derived for momentum. This does not provide very much new information for "ordinary" shallow-water flows and is not given here. The concept is, however, important if there are discontinuities in the flow, such as breaking waves or hydraulic jumps. This is discussed in section 3.4.

3.3.3. Vorticity

Consider (2.51,52) with external forces $(F_x, F_y)/\rho a$ on the right-hand side. You can define a vorticity of the mean flow as

$$\zeta = \frac{\partial v}{\partial x} - \frac{\partial u}{\partial y} \qquad (3.6)$$

This is formally a vector with only the vertical component nonzero, but it can be considered a scalar. Note that this is not the same as the depth mean vorticity. By cross-differentiating (2.51,2.52), the pressure gradient is eliminated and you get an equation for the vorticity:

$$\frac{\partial \zeta}{\partial t} + \frac{\partial}{\partial x}\{u(\zeta + f)\} + \frac{\partial}{\partial y}\{v(\zeta + f)\} + \nabla \times \frac{\tau_b}{\rho a} = \nabla \times \frac{\mathbf{F}}{\rho a} + \nabla \times \mathbf{G} \qquad (3.7)$$

where **G** is a vector with components

$$G_x = \frac{1}{a}\{\frac{\partial}{\partial x}(aT_{xx}) + \frac{\partial}{\partial y}(aT_{xy})\}$$

$$G_y = \frac{1}{a}\{\frac{\partial}{\partial x}(aT_{xy}) + \frac{\partial}{\partial y}(aT_{yy})\}$$

Eq.(3.7) is not formally a conservation law in the sense that it describes fluxes into a column of water. This would require something of the form $\partial/\partial t \, (a\zeta)$. Yet, you may ask the question whether (3.7) says anything on the balance of vorticity. For example, Flokstra (1977) considered steady flow within a *closed* streamline (i.e. **v.n** = 0 on S) without external forcing (i.e. **F** = 0). If you integrate eq.(3.7) over the area within the streamline and apply Stokes' theorem, the result is

$$\iint_A \nabla \times \frac{\tau_b}{\rho a} \, dx \, dy = \int_S \frac{\tau_b}{\rho a} \cdot ds = \int_S \mathbf{G}.ds \qquad (3.8)$$

Along the boundary (being a streamline), the velocity vector must be directed tangentially and therefore the same is true for the bottom stress, so that (with (2.42))

$$\tau_b/\rho = c_f \mathbf{v}|\mathbf{v}| = c_f v_s|v_s|$$

where v_s is the tangential velocity which must have the same sense all around the closed streamline. Then

$$\int_S \frac{\tau_b}{\rho a} \cdot ds = \int_S \frac{c_f}{\rho a}|v_s|^2 ds$$

is either positive or negative definite, depending on the sense of circulation. The only possible source of such a circulation apparently is the right-hand side of (3.8), which involves the lateral stresses (2.19). It was argued by Flokstra that the differential advection parts of the latter usually have the wrong sign, so any driving of the circulation should come from the turbulent stresses (2.46).

3.3.4. Potential vorticity

A sensible conservation law involving vorticity is obtained in a different way. As the Coriolis parameter f does not depend on time, the first term of (3.7) may as well be written as

$$\frac{\partial}{\partial t} (\zeta + f)$$

Performing the differentiation in the advective terms and using (2.48) (also in differentiated form) and finally dividing by a, (3.7) can be written as

$$\frac{\partial}{\partial t}\frac{\zeta+f}{a} + u\frac{\partial}{\partial x}\frac{\zeta+f}{a} + v\frac{\partial}{\partial y}\frac{\zeta+f}{a} = \frac{1}{a}\left(-\nabla\times\frac{\tau_b}{\rho a} + \nabla\times\frac{\mathbf{F}}{\rho a} + \nabla\times\mathbf{G}\right) \qquad (3.9)$$

This says that the *potential vorticity*

$$q = \frac{\zeta+f}{a}$$

will be constant along a streamline if the right-hand side is zero, that is if there are neither external nor frictional forces (Gill, 1982). The potential vorticity plays a very important role in atmospheric and oceanic flows. Note that (3.9) is again not an Eulerian conservation law applying to a control volume fixed in space; neither do you obtain a global balance by integrating over a closed area. It is a Lagrangean conservation law for a fluid volume moving with the flow.

3.3.5. Energy

There is no separate equation for conservation of energy involved in the SWE. You can *derive* one from momentum and mass conservation. Multiply (2.51) and (2.52) by au and av respectively, multiply (2.48) by

$$\tfrac{1}{2}\,(u^2 + v^2) + gh$$

and add them. In order not to complicate things too much, lateral stresses are omitted. Then, realizing that the bottom level does not depend on time, you obtain the following conservation law for energy

$$\tfrac{1}{2}\frac{\partial}{\partial t}\{\,a\,(u^2 + v^2) + g(h^2 - z_b^2)\} + \frac{\partial}{\partial x}\,\{\tfrac{1}{2}au\,(u^2 + v^2)\} + \frac{\partial}{\partial y}\,\{\tfrac{1}{2}av\,(u^2 + v^2)\} +$$
$$+ \frac{\partial}{\partial x}\,(au\,gh) + \frac{\partial}{\partial y}\,(av\,gh) + c_f\,(u^2 + v^2)^{3/2} = F_x u + F_y v \tag{3.10}$$

The quantity

$$\tfrac{1}{2}\{\,a\,(u^2 + v^2) + g(h^2 - z_b^2)\}$$

is the (kinetic + potential) energy per unit surface area. See also Taylor (1919). Note that the contributions of the Coriolis acceleration in the energy equation cancel: as it is directed normal to the flow, no work is done. The external force does produce work on the fluid, which is an energy source. Conversely, the fluid does work by bottom friction, which acts as an energy sink. The other terms describe internal fluxes of kinetic and potential energy. If you integrate (3.10) over a closed area, you see that

$$\tfrac{1}{2}\frac{\partial}{\partial t}\iint_A \{\,a\,(u^2 + v^2) + g(h^2 - z_b^2)\}\,dx\,dy \;=\; \iint_A \{-c_f\,(u^2 + v^2)^{3/2} + F_x u + F_y v\}\,dx\,dy$$

$$\tag{3.11}$$

which says that the change in total energy is due to work by external forces less work by bottom friction. This equation plays a part in the theoretical question of well-posedness of the mathematical problem (see section 5.2).

3.3.6. Enstrophy

Similar to the energy equation you may derive a conservation equation for a quantity indicating the strength of (potential) vorticity. Multiply (3.9) by aq and (2.48) by $q^2/2$ and add them; this gives the conservation law for potential enstrophy

$$\frac{\partial}{\partial t}\,(\tfrac{1}{2}a\,q^2) + \frac{\partial}{\partial x}\,(\tfrac{1}{2}au\,q^2) + \frac{\partial}{\partial y}\,(\tfrac{1}{2}av\,q^2) = q\left(-\nabla\times\frac{\tau_b}{\rho a} + \nabla\times\frac{\mathbf{F}}{\rho a} + \nabla\times\mathbf{G}\right) \tag{3.12}$$

where $aq^2/2$ is defined as the *potential enstrophy*. You see that it is a conserved quantity (in the Eulerian sense) except for external and frictional forces. It is not possible to derive a similar equation for the squared relative vorticity ζ defined as $a\zeta^2/2$.

3.4. Discontinuities

In a few situations, very rapid changes in flow conditions can occur. Examples are hydraulic jumps or tidal waves that get so steep that an almost vertical wave front ("bore") develops. The situation is similar to shock waves in aerodynamics. Viewed on the scale of long waves, such steep fronts are essentially discontinuous. The derivatives in the SWE do not make sense at the fronts and therefore the equations are not valid there. However, the physical ideas of conservation of mass and momentum do still apply; they have just to be cast in a different form.

Consider the mass conservation equation (2.48). If you multiply by some smooth test function ϕ and integrate over the area and over time, you get some sort of weighted mass balance:

$$\int_0^T dt \iint_A \phi \, \{\frac{\partial a}{\partial t} + \frac{\partial}{\partial x}(au) + \frac{\partial}{\partial y}(av)\} \, dx \, dy = 0$$

This may be integrated by parts (Green's theorem) giving:

$$-\int_0^T dt \iint_A \{a\frac{\partial \phi}{\partial t} + au\frac{\partial \phi}{\partial x} + av\frac{\partial \phi}{\partial y}\} \, dx \, dy + \int_0^T dt \int_S \phi \, a \, \mathbf{v}.\mathbf{n} \, ds = 0 \qquad (3.13)$$

where $\mathbf{n}$ is the outward normal on the boundary S and it has been assumed that $\phi = 0$ at $t = 0$ and $t = T$ (this is not essential). Similar integral forms can be constructed from the momentum equations. In these equations, no derivatives of a, u or v occur and therefore, they make sense even if the solution is discontinuous. Such a solution satisfying the integral form of the equations is called a *weak solution*.

Now suppose that the wave (shock) front is a curve in the x-y plane; the curve will move with time and in this way form a curved surface C in x,y,t space (fig. 3.1). First of all, (3.13) is supposed to hold over the entire region, including C. Separately, you may also apply it to a region which just excludes this surface, i.e. you make a cut at C and integrate around it. Then an equation similar to (3.13) results in which both sides of C now occur as boundaries. As no mass is contained within the cut, the mass balance remains perfectly valid:

$$-\int_0^T dt \iint_A \{a\frac{\partial \phi}{\partial t} + au\frac{\partial \phi}{\partial x} + av\frac{\partial \phi}{\partial y}\} \, dx \, dy + \int_0^T dt \int_S \phi \, a \, \mathbf{v}.\mathbf{n} \, ds \, +$$
$$+ \int\int_C \phi \, (a, au, av).\mathbf{n}^* \, dS = 0 \qquad (3.14)$$

Here, $\mathbf{n}^*$ is the normal on C in the 3-d x,y,t space (fig. 3.1). As both (3.13) and (3.14) are valid, their difference must be zero:

$$\int\int_C \phi \, (a, au, av).\mathbf{n}^* \, dS = 0$$

The integral over the two sides of *C* can be split into two integrals along each side. The normal **n*** on one side is just the opposite of the other, so expressing everything in terms of one of the normals, the terms on the other side get a negative sign (remember that ϕ is smooth so it has the same value on both sides). Then you obtain:

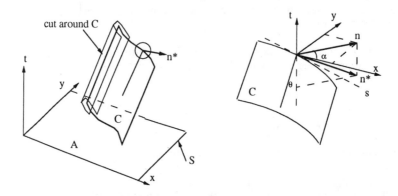

Fig. 3.1. Shock front in *x, y, t* space, with definition of local normal vectors

$$\int\int_C \phi \, [(a, au, av)].\mathbf{n}^* \, dS = 0$$

where [*f*] denotes the "jump" in *f* across the shock surface *C* for any function *f*. As this must be valid for any test function ϕ, the conclusion is that the integrand must vanish at each point of *C*:

$$[(a, au, av)].\mathbf{n}^* = 0$$

Decomposing **n*** into its components (fig. 3.1), this reads

$$- [a] \tan \theta + [au] \cos \alpha + [av] \sin \alpha = 0$$

Here, *tan* θ = *c* is the normal component of the speed of propagation of the shock. Introducing the "usual" normal **n** to the shock in the *x,y* plane, you finally get

$$[a \, \mathbf{v}].\mathbf{n} = [a] \, c \tag{3.15}$$

Physically, this means that the mass flow across the front must be continuous in a frame moving with the shock speed: if there is a jump in water level there must be a corresponding jump in the normal velocity component. The condition (3.15) does not say anything about the tangential velocity component.

The whole reasoning is not restricted to the mass balance; actually, it applies to any conservation law. You may therefore apply the same reasoning to each of the momentum equations. It will be clear that you have to use the conservative form (2.49), (2.50) because the analysis is based on the physical idea of conservation of momentum. Furthermore, you may forget about all driving forces and non-derivative terms (Coriolis, bottom friction) as these will cancel out in the derivation (the volume around C goes to zero). The results can be most clearly represented by switching to a local reference frame s,n parallel and normal to the shock. Then you get from mass conservation (3.15)

$$[av_n] = [a] c \qquad (3.16)$$

for normal momentum

$$[av_n^2 + \tfrac{1}{2}ga^2] = [av_n] c \qquad (3.17)$$

and for tangential momentum

$$[av_n v_s] = [av_s] c \qquad (3.18)$$

From (3.18) and (3.16) you can easily derive that

$$[v_s] = 0 \qquad (3.19)$$

Eqs.(3.16),(3.17) and (3.19) agree with the Rankine-Hugoniot relations for shocks in compressible flow (see the classical book by Courant and Friedrichs, 1948). From the equations, you can express the shock speed c in terms of the jumps in water level and normal velocity. The tangential velocity is continuous; the velocity *difference* between the two sides of the shock is therefore directed normal to the shock.

Chapter 4

Behaviour of solutions

4.1. Linearized equations

If you disregard the lateral viscosity terms, the SWE as given in chapter 2 are of hyperbolic type, that is: they have real characteristics (further discussed below in section 4.3). This implies that you can think of solutions as combinations of waves. The viscosity terms do have an influence: they will act as a damping mechanism. However, they are often not very large and you can generally assume that the wave properties are not essentially altered. This assumption is made in this chapter.

Moreover, the general SWE are nonlinear and although it is possible to analyze the behaviour of the solutions, e.g. in terms of characteristics, most of the relevant properties are more easily obtained by considering the linearized case. In this book, this approach is used to analyse the properties of solutions of the SWE in continuous form. Experience shows that in many cases a linear analysis even gives quantitatively correct results in order of magnitude, that is, the flows are often not too far from linear. In later chapters, the performance of numerical methods for these linear solutions is investigated. A necessary condition for numerical methods is that they should work satisfactorily at least for linear waves, even though this does not imply that they will do so for the nonlinear phenomena as well. Just to make things perfectly clear, I do *not* propose to use the linearized and simplified equations for practical application, they serve only to understand what is going on.

To linearize the SWE, assume that you have small perturbations (indicated by a prime) on a steady uniform flow:

$$u = U + u'$$
$$v = V + v'$$
$$a = a_0 + a'$$

Introducing this into the SWE (2.48), (2.51), (2.52) you can collect all terms of equal order. You will find that the zero-order terms cancel because it is assumed that the reference state satisfies the SWE. Then neglecting all terms involving higher than first powers of the perturbations, you get the linear approximation

$$\frac{\partial u}{\partial t} + U\frac{\partial u}{\partial x} + V\frac{\partial u}{\partial y} - fv + g\frac{\partial h}{\partial x} + ru = 0$$

$$\frac{\partial v}{\partial t} + U\frac{\partial v}{\partial x} + V\frac{\partial v}{\partial y} + fu + g\frac{\partial h}{\partial y} + rv = 0 \tag{4.1}$$

$$\frac{\partial h}{\partial t} + U\frac{\partial h}{\partial x} + V\frac{\partial h}{\partial y} + a_0(\frac{\partial u}{\partial x} + \frac{\partial v}{\partial y}) = 0$$

The primes on the variables have been omitted. It is understood that U, V and a_0 are constants. Here, an approximation has been made in the bottom friction terms in order not too complicate things too much. Formally,

$$c_f\frac{(U + u')\sqrt{(U + u')^2 + (V + v')^2}}{a_0 + a'} = c_f\frac{U\sqrt{U^2 + V^2}}{a_0} + r_1u' + r_2v' + r_3a'$$

In (4.1), the terms with v' and a' have been neglected under the assumption that the main effect of friction is represented by the term r_1u, where r_1 is a linearized frictional coefficient proportional to c_f. The subscript 1 has been omitted in (4.1). This approximation only leads to some inconsistency on the case of flood waves (section 4.4.3) but the qualitative behaviour is correct.

This way of linearizing does not work very well for a situation which occurs rather often in practice: oscillating (e.g. tidal) flow on a *small* steady flow. Then u' is not small relative to U and the linearization is not valid: the entire friction term would be second order and vanish in the linearized equations. Yet, it is known that bottom friction does have a non-negligible influence in such cases, so we must find a simplified representation to take it into account. Such an approach has been given in a little-known paper by Lorentz (1937). An artificial linear bottom friction is introduced which dissipates the same amount of energy as a quadratic one would do during a complete tidal cycle. For 1-d flow, the work done during a tidal cycle is the product of bottom stress and velocity:

$$\int_0^T u\,\tau_b/\rho\, dt = \int_0^T c_f u^2|u|\, dt$$

if a quadratic friction is used and

$$\int_0^T \lambda u^2\, dt$$

if a linearized friction law $\tau_b/\rho = \lambda u$ is used (note that λ has the dimension of velocity). Assuming a periodic flow $u = u_0 \sin \omega t$ and equating the two expressions, the frictional coefficient λ can be solved:

$$\lambda = \frac{8}{3\pi}c_f u_0 \quad \text{so} \quad r = \frac{\lambda}{a_0} = \frac{8}{3\pi a_0}c_f u_0 \tag{4.2}$$

(compare with the discussion of tidal stresses in section 2.6). This requires that you can estimate the magnitude of the oscillating component beforehand. For the 2-d case, you generally find anisotropy: the coefficients in x- and y-directions are different. However, as a first approximation we use the same coefficient in both directions.

In oceanography, there is a different motivation for a linear bottom friction. In a rotating flow, a viscous boundary layer at the bottom (Ekman layer) will be formed, the net effect of which on the inviscid interior flow is given by such a linear expression (see, e.g., Pedlosky, 1979).

For later use, (4.1) can conveniently be written in vector-matrix form:

$$\frac{\partial \mathbf{v}}{\partial t} + A \frac{\partial \mathbf{v}}{\partial x} + B \frac{\partial \mathbf{v}}{\partial y} + C \, \mathbf{v} = 0 \tag{4.3}$$

where

$$\mathbf{v} = \begin{pmatrix} u \\ v \\ h \end{pmatrix} \qquad A = \begin{pmatrix} U & 0 & g \\ 0 & U & 0 \\ a & 0 & U \end{pmatrix} \qquad B = \begin{pmatrix} V & 0 & 0 \\ 0 & V & g \\ 0 & a & V \end{pmatrix} \qquad C = \begin{pmatrix} r & -f & 0 \\ f & r & 0 \\ 0 & 0 & 0 \end{pmatrix}$$

The subscript 0 on a will be dropped if there is no danger of confusion.

The potential-vorticity equation (3.9) can be linearized in the same way. You will have to take into account that the Coriolis parameter f may depend on geographical latitude; the North direction is then indicated traditionally by y. The β-plane approximation is often used in which f varies linearly with y :

$$f = f_0 + \beta y$$

The linearized form of potential vorticity becomes

$$\frac{\zeta + f}{a} = \frac{f_0 + \beta y + \zeta'}{a_0} - \frac{f_0 a'}{a_0^2} = q_0 + q$$

where the reference and perturbation potential vorticities are

$$q_0 = \frac{f_0 + \beta y}{a_0} \qquad q' = \frac{\zeta'}{a_0} - \frac{f_0 a'}{a_0^2}$$

Eq.(3.9), without viscous or external terms, is then linearized as

$$\frac{\partial q'}{\partial t} + U \frac{\partial q'}{\partial x} + V \frac{\partial q'}{\partial y} + \frac{\beta}{a_0} v' + \frac{r}{a_0} \zeta' = 0 \tag{4.4}$$

You could have derived this directly from (4.1) as well. The subscript 0 on f will be dropped if there is no danger of confusion.

To simplify things even further, you can transform all equations in this section to a reference frame moving with the mean flow (U, V). The effect is that the advection terms disappear. However, in some cases (e.g. the analysis of boundary conditions) the flow direction is important and you have to take it essentially into account.

4.2. Wave equation

A single wave equation for the water level can be obtained if you eliminate the velocity variables. This is easy to see you omit for a while the advective, Coriolis and friction terms from (4.1). Taking the time derivative of the continuity equation and substituting the momentum equations:

$$\frac{\partial^2 h}{\partial t^2} - ga\left(\frac{\partial^2 h}{\partial x^2} + \frac{\partial^2 h}{\partial y^2}\right) = 0 \tag{4.5}$$

which is the standard wave equation. Lynch and Gray (1979) have used this equation in a numerical method arguing that it has certain numerical advantages over the continuity equation (4.1c) (see also Kinnmark, 1986).

Applying the same derivation to the complete equations, the resulting equation still contains velocity variables. You can see the full form in the references; here it is sufficient to write down the linearized form. Actually, they multiply (4.1a,b) by depth a and subtract not just the time derivative of (4.1c), but $(\partial/\partial t + G)$ times the equation, with a parameter G to be decided upon. In a moving reference frame you will find

$$\frac{\partial^2 h}{\partial t^2} - ga\left(\frac{\partial^2 h}{\partial x^2} + \frac{\partial^2 h}{\partial y^2}\right) + G\frac{\partial h}{\partial t} + (G-r)a\left(\frac{\partial u}{\partial x} + \frac{\partial v}{\partial y}\right) + af\left(\frac{\partial v}{\partial x} - \frac{\partial u}{\partial y}\right) - au\beta = 0 \tag{4.6}$$

This equation would be used instead of the original equation of continuity (4.1c). Because (4.6) is of higher order than (4.1c), a spurious solution is introduced. Kinnmark (1986) shows that it is damped approximately as exp($- Gt$) and therefore he recommends not to take $G = 0$ but rather $G = r$ (this is further discussed in section 4.4). It is possible in principle to eliminate the velocity variables completely to obtain a general wave equation with water level h as the only dependent variable (for a similar derivation see section 4.4.4). This has not been done in the references just cited and it is not known whether it results in numerical advantages.

4.3. Characteristics

Although the theory of characteristics can be formulated for quasilinear equations such as the SWE, in this book only the linear form is given as it provides sufficient information for the numerical analysis in later chapters. For an extensive theory of characteristics see Courant and Hilbert (1962).

A somewhat loose, heuristic definition of a characteristic surface in x,y,t space is that it is a surface on which the set of equations (4.1) can be formulated in terms of only two independent variables. For this purpose only the derivative terms need to be considered.

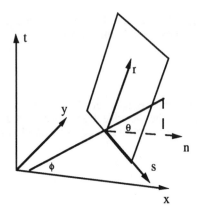

Fig. 4.1. Definitions for a characteristic plane in (x,y,t) space

The equations are transformed to a system of coordinates r,n,s, as shown in fig. 4.1 where n is normal to the assumed characteristic surface and r and s are in the surface. If an observer travels along the r- axis, his propagation speed c is related to the angle θ by $\tan \theta = 1/c$. The coordinate transformation is given by

$$
\begin{pmatrix} \dfrac{\partial}{\partial t} \\[2mm] \dfrac{\partial}{\partial x} \\[2mm] \dfrac{\partial}{\partial y} \end{pmatrix} = \begin{pmatrix} -\sin \theta & \cos \theta & 0 \\[1mm] \cos \theta \cos \phi & \sin \theta \cos \phi & -\sin \phi \\[1mm] \cos \theta \sin \phi & \sin \theta \sin \phi & \cos \phi \end{pmatrix} \begin{pmatrix} \dfrac{\partial}{\partial n} \\[2mm] \dfrac{\partial}{\partial r} \\[2mm] \dfrac{\partial}{\partial s} \end{pmatrix}
$$

and application to (4.3) gives

$$
(-\sin \theta\, I + \cos \theta \cos \phi\, A + \cos \theta \sin \phi\, B) \frac{\partial \mathbf{v}}{\partial n} +
$$

$$
+ (\cos \theta\, I + \sin \theta \cos \phi\, A + \sin \theta \sin \phi\, B) \frac{\partial \mathbf{v}}{\partial r} + (-\sin \phi\, A + \cos \phi\, B) \frac{\partial \mathbf{v}}{\partial s} = 0
$$

$$
\tag{4.7}
$$

where I is the unit matrix. If the plane is to be characteristic, a derivative with respect to n should not occur, which means that

$$
(-\sin \theta\, I + \cos \theta \cos \phi\, A + \cos \theta \sin \phi\, B) \frac{\partial \mathbf{v}}{\partial n} = 0
$$

This is not generally possible. The homogeneous equations have nontrivial solutions (eigenvectors) only if the determinant is zero:

$$\det(-\tan\theta\, I + \cos\phi\, A + \sin\phi\, B) = \tag{4.8}$$

$$= \begin{vmatrix} u_n\text{-}c & 0 & g\cos\phi \\ 0 & u_n\text{-}c & g\sin\phi \\ a\cos\phi & a\sin\phi & u_n\text{-}c \end{vmatrix} = -(c-u_n)\{(c-u_n)^2 - ga\} = 0$$

where $u_n = U\cos\phi + V\sin\phi$ is the velocity component normal to the s-axis This means that there are three possible slopes (eigenvalues) of the characteristic surface for any orientation ϕ:

$$c_1 = u_n \tag{4.9}$$
$$c_{2,3} = u_n \pm \sqrt{ga}$$

The corresponding eigenvectors are

$$\mathbf{e}_1 = \begin{pmatrix} \sin\phi \\ -\cos\phi \\ 0 \end{pmatrix} \qquad \mathbf{e}_{2,3} = \begin{pmatrix} \pm\sqrt{g/a}\,\cos\phi \\ \pm\sqrt{g/a}\,\sin\phi \\ 1 \end{pmatrix}$$

and a general field will be a superposition of such contributions. Still restricting yourself to one direction ϕ, this gives

$$\mathbf{v} = \sum_{j=1}^{3} \alpha_j(r,n,s)\mathbf{e}_j$$

and (4.7) shows that the α's satisfy

$$(\cos\theta\, I + \sin\theta\cos\phi\, A + \sin\theta\sin\phi\, B)\frac{\partial\alpha_j}{\partial r} + (-\sin\phi\, A + \cos\phi\, B)\frac{\partial\alpha_j}{\partial s} = 0$$

This scalar equation describes disturbances propagating in characteristic surfaces only; you may consider these (loosely) as waves. If you have a wave with its crest in the s-direction, it can apparently propagate at three speeds in a direction normal to itself, as given by (4.9). The fact that you find three distinct real directions with independent eigenvectors, corresponding to the order of the system of differential equations, defines the system to be *hyperbolic*. However, this is no longer so if the horizontal diffusion terms of (2.46) are taken into account which introduce a parabolic character. As they occur only in the momentum equations and not in the continuity equation, the system has been called incompletely parabolic in that case (Sundström, 1977). The diffusion terms are generally considered to be relatively small corrections so the system is mainly treated as hyperbolic.

In general, you will have a superposition of many waves with different orientations. A number of these are shown in fig. 4.2 and they form two families. There is a family of waves moving at speed c_1. According to (4.9), particles moving at velocity U, V (in the original x, y, t frame) stay in the characteristic plane.

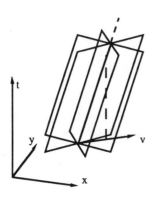 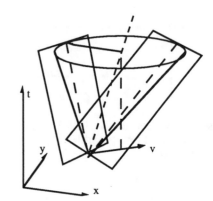

Fig. 4.2. Manifolds of characteristic planes

All planes, irrespective of the orientation, have the particle path in common (fig. 4.2a) so they form a manifold of planes with the particle path as its axis. This type of waves does not have a parallel in 1-d wave propagation.

The other type of waves is also known for 1-d shallow-water flow: the waves move at speeds $c_{2,3}$ and the characteristic planes are tangent to an oblique cone with the particle path as its axis and its top angle determined by the signal speed $\sqrt{ga}$ (fig. 4.2b). If the magnitude of the flow velocity is such that

$$|v| < \sqrt{ga} \quad \text{or} \quad F < 1 \tag{4.10}$$

where F is the Froude number (section 3.1), the waves can propagate in all directions; the flow is then called *subcritical* (the aerodynamic equivalent is subsonic). In the opposite case, the waves propagate only in a certain conical region; this situation is, however, not very common in hydrology.

The theory of characteristics is difficult to handle in more general situations. It is important to determine the requirements for boundary conditions (chapter 5). In this book a different approach is preferred. In section 4.4 all previous results from characteristic theory are confirmed by an analysis of harmonic waves; moreover, that analysis gives more information on the influence of the non-derivative terms (Coriolis, bottom friction), which may be considerable.

In a *steady state*, you have a problem in two spatial dimensions. You might wonder whether any characteristics exist in the x-y plane; these would be lines instead of surfaces. To find this out, you follow the same reasoning and try to find a curve along which only derivatives in the tangential direction s are left and the derivatives in the normal direction n cancel. Suppose the angle between the curve and the x-axis to be θ, then the equations (4.3) would take the form

$$(A \sin \theta - B\cos \theta) \frac{\partial \mathbf{v}}{\partial n} + (A \cos \theta + B \sin \theta) \frac{\partial \mathbf{v}}{\partial s} = 0$$

from which the normal derivatives cancel if the determinant vanishes and $\partial v/\partial n$ is proportional to an eigenvector:

$$det\,(\lambda A - B) = \begin{vmatrix} \lambda U - V & 0 & \lambda g \\ 0 & \lambda U - V & g \\ \lambda a & a & \lambda U - V \end{vmatrix} =$$

$$= (\lambda U - V)\{(\lambda U - V)^2 - ga\} - (\lambda U - V)\,\lambda^2 ga = 0 \qquad (4.11)$$

with $tan\,\theta = \lambda$. Eq.(4.11) has three roots again. The first is

$$\lambda U - V = 0 \qquad (4.12)$$

which describes a curve tangent to the velocity vector, that is a particle path. The other two roots satisfy

$$\left(\lambda U - V\right)^2 - ga\,(\lambda^2 + 1) = (U^2 - ga)\,\lambda^2 + 2UV\lambda + V^2 - ga = 0 \qquad (4.13)$$

If the discriminant of this equation for λ is negative, there are no real roots and consequently no characteristics. This is the case if

$$U^2 + V^2 - ga < 0$$

which you recognize as the condition for subcritical flow (4.10). In this case, there are less than 3 real eigenvalues so the (steady-state) system is not hyperbolic but elliptic. If the flow is supercritical, there are two real characteristics

$$\lambda = \frac{UV \pm \sqrt{ga(U^2 + V^2 - ga)}}{U^2 - ga} \qquad (4.14)$$

which correspond to Mach lines in aerodynamics. Both cases can be recognized in (4.5), which in fixed coordinates becomes (for the special case $V = 0$ as an illustration)

$$(U^2 - ga)\,h_{xx} - ga\,h_{yy} = 0$$

which is indeed a hyperbolic equation if the first coefficient is positive and otherwise elliptic. As mentioned before, the former case is not very common in shallow-water flows.

4.4. Harmonic wave propagation

In the analysis of characteristics, the non-derivative terms in the SWE (bottom friction, Coriolis acceleration) have not been taken into account, in agreement with the mathematical theory. These terms do not influence the *type* of the equations. However, they can have an essential influence on the behaviour of solutions. Most of the solutions have a wave character as already found in the theory of characteristics, but they may behave quite differently depending on the rotation or friction parameters. For the linearized equations, a harmonic wave analysis (including the lower order terms) gives just the insight into the behaviour of the waves that is not so easily produced by the theory of characteristics.

Assume a wave of the form

$$\mathbf{v} = \mathbf{v}_1 \, exp\{\, i(k_1x + k_2y - \omega t)\}$$

with wave number (k_1, k_2) fixed, (complex) frequency ω and (complex) amplitude vector $\mathbf{v}_1$ to be determined. If ω has a nonzero imaginary part, this corresponds to wave damping or amplification. For the time being, suppose that the Coriolis parameter f is a constant. Substituting into (4.3), you get

$$(- i\omega I + ik_1A + ik_2B + C)\mathbf{v}_1 = \begin{pmatrix} r - i\omega & -f & ik_1g \\ f & r - i\omega & ik_2g \\ ik_1a & ik_2a & -i\omega \end{pmatrix} \begin{pmatrix} u_1 \\ v_1 \\ h_1 \end{pmatrix} = 0 \tag{4.15}$$

This homogeneous system of equations has a nontrivial solution only if its determinant vanishes, which gives

$$- i\omega\{(r - i\omega)^2 + f^2\} + ga\, k^2 \, (r - i\omega) = 0 \tag{4.16}$$

with $k^2 = k_1^2 + k_2^2$. There are three roots (eigenvalues) for ω to this equation, which correspond to three waves, all having the same wave number (i.e. wave length and direction) but possibly different phase speeds, damping rates and amplitudes. The latter follow from the eigenvectors that can be obtained from (4.15) once you have the eigenvalues from (4.16). It is not easy to solve (4.16) in the general case. However, it is very instructive to solve some special cases which bring out the various types of waves.

4.4.1. Gravity waves

First assume f to be small; then (4.16) has a root $\omega_1 = - ir$ to be discussed in the next section, and two roots from

$$- \omega^2 - i\omega r + ga\, k^2 = 0 \tag{4.17}$$

If, furthermore, you consider the frictionless case ($r = 0$), the equation becomes

$$\omega^2 - ga\, k^2 = 0$$

Translating this back to the original fixed reference frame, this would read

$$(\omega - k\, u_n)^2 - ga\, k^2 = 0 \tag{4.18}$$

where u_n is the velocity component along the wave number vector, that is normal to the wave front:

$$u_n = (k_1U + k_2V)/k = U \cos \phi + V \sin \phi$$

and you recognize (4.18) as the equation for characteristic planes (4.9) if you realize that the phase speed (in a direction normal to the wave front) is $c = \omega/k$. This confirms that the present analysis gives the same results as the theory of characteristics if there are no lower-order terms. Also, the wavelike behaviour of the solutions is more clearly brought out.

For this case, the eigenvectors can be found from (4.15):

$$\mathbf{v}_1 = (\pm\sqrt{g/a}\ \cos\phi, \pm\sqrt{g/a}\ \sin\phi,\ 1)^T \qquad (4.19)$$

again in agreement with the theory of characteristics. The vector of velocity variation is directed parallel to the wave number vector (normal to the wave front), so you could call this a *longtudinal wave*. The water-level variation and velocity variations are in phase, and the magnitude of the water level variation is related to the velocity variation according to the ratio $1 : \sqrt{g/a}$.

You could call waves with these properties *pure gravity waves*; in a moving reference frame they are described by the "model" equations

$$\frac{\partial u}{\partial t} + g\frac{\partial h}{\partial x} = 0$$
$$\frac{\partial v}{\partial t} + g\frac{\partial h}{\partial y} = 0 \qquad (4.20)$$
$$\frac{\partial h}{\partial t} + a\left(\frac{\partial u}{\partial x} + \frac{\partial v}{\partial y}\right) = 0$$

which could be called the basic form of the shallow-water equations.

In many cases, bottom friction has only a relatively small influence on wave propagation, but there are situations where it gets large, even dominant (see section 4.4.3).

If the Coriolis coefficient f gets more important, you find a modification of the gravity waves. For simplicity, assume bottom friction still to be small; then (4.16) has one root $\omega_1 = 0$ (to be discussed in the next section) and two roots

$$\omega_{2,3} = \pm\sqrt{f^2 + ga\,k^2} \qquad (4.21)$$

For relatively short waves (large k), this reduces to the previous case. In the other exetreme, for very long waves (small k), $\omega = \pm f$; these are called inertial waves, the period of which is related to the earths' rotation rate; it varies from 12 h at the poles to several days at lower latitudes. This case occurs if the term gak^2 in (4.21) is small, so the corresponding wave lengths satisfy

$$L \gg 2\pi\frac{\sqrt{ga}}{f} = 2\pi R$$

where $R = \sqrt{ga}\ /f$ is the *Rossby radius of deformation*. Its value is several thousand km in the ocean, but several hundred km in shallow water, such as the North Sea. Moreover, for internal modes (section 2.5.2) the equivalent water depth may be of the order of just 1 m, with a corresponding internal Rossby radius of several tens of km.

For the general case of finite f and k, the phase speed $c = \omega/k$ from (4.21) is larger than that of pure gravity waves, as shown in fig. 4.3. The Coriolis term now has a certain influence and the gravity waves are called *Poincaré waves*. Note that this influence is not found from the theory of characteristics. The eigenvectors can be derived as for pure gravity waves, but they are not shown here. The equations describing frictionless Poincaré waves (in a reference frame moving with the mean flow U,V) are

$$\frac{\partial u}{\partial t} - fv + g\frac{\partial h}{\partial x} = 0$$

$$\frac{\partial v}{\partial t} + fu + g\frac{\partial h}{\partial y} = 0 \qquad (4.22)$$

$$\frac{\partial h}{\partial t} + a(\frac{\partial u}{\partial x} + \frac{\partial v}{\partial y}) = 0$$

Unlike pure gravity waves, Poincaré waves are dispersive (the phase speed depends on $\mathbf{k}$) and consequently they have a group velocity $\mathbf{c_g}$ deviating from the phase speed c. The components of the group velocity vector are

$$c_{gj} = \frac{\partial \omega}{\partial k_j} = \frac{\sqrt{ga}\, k_j}{\sqrt{k^2 + f^2/ga}} \qquad j = 1,2$$

The vector is directed normal to the wave front and its magnitude is less than the phase speed (see fig. 4.3).

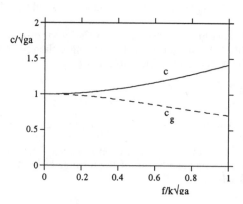

Fig. 4.3. Phase and group velocity of Poincaré waves

4.4.2. Potential vorticity waves

Eq. (4.16) has always one small root. We consider two cases. If there is friction but no rotation ($f = 0$, $r \neq 0$), you find a root $\omega_1 = - ir$. It corresponds to an exponentially damped wave at a rate $\exp(-rt)$. Its real part is zero, so the wave is stationary in the moving reference frame, i.e. it moves with the mean flow in a fixed frame. This corresponds to the "slow" characteristic c_1 from (4.9). The eigenvector is

$$\mathbf{v}_1 = (-\sin \phi, \cos \phi, 0)^T$$

which says that the velocity variation is now *parallel* to the wave front. This is therefore a *transversal* wave. Furthermore, there is no water-level variation involved and consequently, you may take $a' = 0$ in the potential-vorticity equation (4.4) which then reduces to simple damping (and advection in a fixed reference frame) of vorticity (which is just proportional to potential vorticity in this case)

$$\frac{\partial \zeta}{\partial t} + r \, \zeta = 0 \tag{4.23}$$

In the other case, where $f \neq 0$ (but still constant) and friction is absent, you get a root $\omega_1 = 0$ from (4.16) which again corresponds to a wave moving with the mean flow, but now without damping. From (4.15), you can reconstruct the governing equations for this case:

$$
\begin{aligned}
- fv + g \frac{\partial h}{\partial x} &= 0 \\
fu + g \frac{\partial h}{\partial y} &= 0 \\
\frac{\partial u}{\partial x} + \frac{\partial v}{\partial y} &= 0
\end{aligned}
\tag{4.24}
$$

The first two equations state that the flow is in geostrophic equilibrium. However, this case is degenerate as any geostrophic flow satisfies the third equation and therefore is not completely specified. The potential-vorticity equation (4.4) gives

$$\frac{\partial q}{\partial t} = 0$$

which says that the initial vorticity is conserved (moving with the mean flow). This, together with (4.24a,b) and the definition of q provides a set of equations for such waves.

4.4.3. Flood waves

In (4.17), the frictional coefficient r may have a dominant influence. The two roots

$$\omega_{2,3} = - \frac{ir}{2} \pm \sqrt{gak^2 - r^2/4}$$

are shown in fig. 4.4 as a function of the parameter gak^2/r^2. For small values of this parameter (r large and/or k small), one of the roots approaches

$$\omega_2 \approx -ir$$

Then there are two such roots (the other one is discussed in section 4.4.2). The third root approaches

$$\omega_3 = -i\frac{ga\, k^2}{r} \tag{4.25}$$

This describes a wave slowly damped at a rate

$$exp(-\frac{ga\, k^2}{r}\, t)$$

The real part of ω_3 is zero, indicating that the wave is stationary in the moving reference frame, i.e. moving with the mean flow velocity. The approximation is valid if

$$4gak^2/r^2 \ll 1 \quad \text{or} \quad L \gg 4\,\pi\sqrt{ga}/r$$

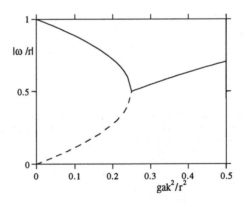

Fig. 4.4. Frequency of waves influenced by bottom friction

which enables you to estimate the wave lengths involved. The other two waves having $\omega = -ir$ are damped at a rate $exp(-rt)$ which is very fast in comparison with that of the third wave. For practical purposes, you can assume that they do not survive.

This is a representation of what is known in river engineering as *flood waves*. These are very long waves, originating from, e.g., rainfall periods of several days. You should realize that a flood wave is not a new type of wave, but it just one of the three waves inherent in the SWE, in the special case of large frictional influence. There is a continuous transition from the pure gravity-wave case to the present case of flood waves, with increasing frictional effect. In the intermediate range, you find gravity waves significantly influenced by friction.

The description is not quite accurate quantitatively due to the neglect of parts of the linearized friction terms (see the discussion of (4.1)). If you had taken all contributions into account, you would have obtained a qualitatively similar but more correct result. In particular, the speed of propagation would turn out to be not just the flow velocity but about 1.5 times larger; this depends somewhat upon the exact formulation of the bottom friction term (for 1-d wave propagation see Vreugdenhil, 1989). However, if you keep this in mind, (4.25) is good enough for the present purpose.

The eigenvector follows from (4.15) again; if you neglect $i\omega$ in comparison with r (which is allowed in view of (4.25) with the assumption that gak^2/r^2 is small), you get

$$\mathbf{v}_1 = (cos\ \phi,\ sin\ \phi,\ -ir/gk)^T$$

The velocity variation is again in the direction of wave propagation; the water level variation is out of phase (indicated by the factor i). The resulting simplified set of equations is

$$ru + g\frac{\partial h}{\partial x} = 0$$

$$rv + g\frac{\partial h}{\partial y} = 0 \qquad (4.26)$$

$$\frac{\partial h}{\partial t} + a\left(\frac{\partial u}{\partial x} + \frac{\partial v}{\partial y}\right) = 0$$

The flow turns out to be in quasi-steady equilibrium; the only time-dependent term is the water-level variation in the continuity equation. From (4.26), you can eliminate the velocity components to find a diffusion equation for the water level:

$$\frac{\partial h}{\partial t} - \frac{ag}{r}\left(\frac{\partial^2 h}{\partial x^2} + \frac{\partial^2 h}{\partial y^2}\right) = 0 \qquad (4.27)$$

For this reason, this approximation is known as the *diffusion analogy* for flood waves.

4.4.4. Rossby waves

The variation of the Coriolis parameter f with geographical latitude may significantly change the behaviour of vorticity waves. Let us start from (4.22) but now assume that f is a function of y. You can follow a somewhat different route than in the preceding sections by eliminating two of the unknowns. From (4.22a,c), u and h can be expressed in terms of v:

$$h_{tt} - ga\ h_{xx} = -afv_x - a\ v_{yt}$$
$$u_{tt} - ga\ u_{xx} = fv_t + ga\ v_{xy}$$

Apply the same operation as in the left hand side of these equations to (4.22b) and substitute the expressions for u and h ; then you obtain one equation with one unknown

$$v_{ttt} - ga \ (v_{xx} + v_{yy}) + f^2 v_t - ga\beta \ v_x = 0 \qquad (4.28)$$

where $\beta = df/dy$. In (4.28) the main effect of the variation of f is incorporated in the β-term, so as an approximation you might now suppose f to be constant in the other terms. This is allowed if the spatial scale is not too large ($\beta L \ll f$ or $\beta \ll fk$). Then (4.28) has constant coefficients and you can assume harmonic solutions again. The dispersion relation turns out to be

$$\omega^3 - (ga \ k^2 + f^2)\omega - ga \ \beta \ k_1 = 0 \qquad (4.29)$$

This equation can be solved exactly, but it is more instructive to take into account the orders of magnitude. Under the given assumption on the magnitude of β, (4.29) apparently has one small and two large roots:

$$\omega_1 \approx -\frac{\beta \ ga \ k_1}{ga \ k^2 + f^2} \qquad (4.30)$$

$$\omega_{2,3} \approx \pm\sqrt{ga \ k^2 + f^2}$$

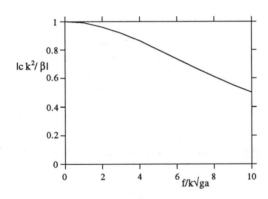

Fig. 4.5. Phase speed of Rossby waves

The latter are identical with the Poincaré waves (4.21). The former is of the same variety as the vorticity wave (section 4.4.2); however, it does not just move with the mean flow but is significantly influenced by the differential rotation β. It is known as *Rossby wave*. From (4.30) a phase velocity $c_1 = \omega_1/k_1$ in x-direction can be deduced. Its value is always negative, so Rossby waves are moving Westward relative to the mean flow. The phase speed is illustrated in fig. 4.5. Contours of constant ω in the **k** plane are circles as shown in fig. 4.6. The center is at

$$k_1 = -\frac{\beta}{2\omega}$$

and the radius is

$$\left\{ \left(\frac{\beta}{2\omega}\right)^2 - \frac{f^2}{ga} \right\}^{1/2}$$

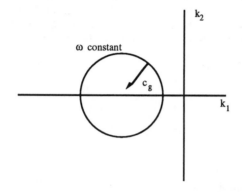

Fig. 4.6. Contour of constant ω in wave-number space for Rossby waves, with group velocity

Rossby waves, too, are dispersive. The components of the group velocity vector are found to be

$$c_{gx} = \frac{\beta}{(k^2 + f^2/ga)^2} (k_1^2 - k_2^2 - \frac{f^2}{ga})$$

$$c_{gy} = \frac{\beta}{(k^2 + f^2/ga)^2} 2k_1 k_2$$

(4.31)

Because of its definition, the group velocity has a direction perpendicular to the circle in fig. 4.6. This means that it can have a component either Westward or Eastward; the same is then true for the energy flux.

The eigenvector for Rossby waves is found to be

$$\mathbf{v}_1 = (- \sin \phi, \cos \phi, \frac{-if}{gk})^T$$

which indicates that the wave is transversal like the vorticity wave. Unlike (4.19) there is now a nonzero water-level variation; it is out of phase with the velocity and the ratio to the velocity variation is f/gk. In the case of gravity waves (4.19) this was $\sqrt{a}/g$; so compared with that case, the water-level variations are now smaller by a factor

$$\frac{f}{k\sqrt{ga}}$$

This is small for wave lengths small compared with the deformation radius, so in that case the water level is approximately undisturbed as in the case of vorticity waves.

It is not so easy to reconstruct the model equations for a Rossby wave. An extensive discussion is given by Gill (1982) or Pedlosky (1979). They conclude that departures from geostrophic equilibrium are important for the dynamics. The potential-vorticity equation is derived including local acceleration terms (4.32c below). Next, if the length scales are limited (as indicated above), the momentum equations can be simplified to a local geostrophic equilibrium. Together this constitutes the approximate system of equations for Rossby waves

$$-fv + g\frac{\partial h}{\partial x} = 0$$

$$fu + g\frac{\partial h}{\partial y} = 0 \tag{4.32}$$

$$\frac{\partial q}{\partial t} + \beta\frac{v}{a} = 0$$

with all coefficients now taken constant and

$$q = \frac{1}{a}(\frac{\partial v}{\partial x} - \frac{\partial u}{\partial y}) - \frac{fh}{a^2}$$

You can check that this system has a travelling wave solution identical to (4.30).

4.4.5. Wave equation

Harmonic wave propagation can also be applied to the wave equation (4.6) in combination with the momentum equations (4.1a,b). For the special case $f = 0$, you will find that there is a root $\omega = -ir$, which is the vorticity wave; next a cubic equation is found:

$$i\omega^3 - (r + G)\omega^2 - i\omega (Gr + ga\, k^2) + ga\, k^2 G = 0$$

Its order is higher than in section 4.4.1. which reflects the fact that a spurious, non-physical solution has been introduced. You may check that the cubic equation always has a root $\omega = -iG$ so the spurious solution is a stationary mode damped at a rate $\exp(-Gt)$. If $G = 0$, it is obvious that this mode is undamped, which may be unpleasant because, if excited in some way, it will not decay. For this reason, Kinnmark (1986) proposes to take $G \neq 0$, e.g. $G = r$. The other roots follow from

$$-\omega^2 - ir\,\omega + ga\, k^2 = 0$$

which is identical with (4.17).

Chapter 5

Boundary conditions

5.1. Characteristic arguments

A set of differential equations does not mean anything unless appropriate boundary conditions are specified. The number of boundary (or initial) conditions is closely related to the behaviour of characteristics if the system is hyperbolic. The SWE are hyperbolic if you do not take lateral friction into account. You can find the relevant theory in Courant & Hilbert (1962). It says first of all that the region of interest should be enclosed by a boundary with no holes in it. Such boundaries can obviously be coastlines, but if you take a region covering part of the sea, you also have an "open" boundary which is artificial in the sense that it is not a physical boundary: it is just a line drawn on the map.

Second, at any point of the boundary, you should specify *as many boundary conditions as there are characteristic planes entering the region.* The speeds of propagation are given in (4.9). The orientation of the wave front is arbitrary, so you have to choose one and decide which of the three characteristic planes is propagating into the region. The same information can be obtained from the wave interpretation of section 4.4 which shows that waves are moving with the flow velocity plus a phase speed defined by, e.g. (4.17). In both cases you see that it is the velocity component of the base flow (reference field) *normal to the boundary* which determines whether waves are moving in or out. We indicate this by U_n. Counting it positive if the flow is outward, you can have the following situations:

	condition	number of outward char.	number of inward char.	number of boundary conditions
1.	$U_n < -\sqrt{ga}$	0	3	3
2.	$-\sqrt{ga} \leq U_n < 0$	1	2	2
3.	$0 \leq U_n < \sqrt{ga}$	2	1	1
4.	$U_n \geq \sqrt{ga}$	3	0	0

One of the conclusions is that the number of boundary conditions is not fixed. The first and last cases are concerned with supercritical flow and do not occur very often. Nevertheless, it is theoretically conceivable to have supercritical inflow at all boundaries and therefore three boundary conditions all around the boundary, or (as the other extreme) supercritical outflow at all boundaries and no boundary conditions whatsoever.

The normal cases in practice, however, are cases 2 and 3 and you conclude that you must specify either one or two boundary conditions, depending on whether the flow is directed outward or inward. This situation may change in time and it normally does so in tidal flow. On a coastline, the normal velocity is zero and you need only one condition, which is exactly that the normal velocity is zero.

The *type* of boundary conditions to be specified has something to do with the physical significance of the characteristics, that is, what is transported along them. This has been studied in section 4.4 in terms of waves. The theory of Courant & Hilbert can be explained such that you have to specify any wave that comes in from outside. The "fast" characteristics (2 and 3 in (4.9)) correspond to gravity waves. You can provide information about them by prescribing either

$$h = f(s,t) \qquad\qquad (5.1)$$

or $\quad u_n = g(s,t) \qquad\qquad (5.2)$

(where s is a coordinate along the boundary and f and g are functions describing the incoming waves in some way) or a combination of both, e.g. the discharge per unit width across the boundary

$$a\, u_n = q(s,t) \qquad\qquad (5.3)$$

You should be aware that such boundary conditions will lead to (partial) reflection if there is any outgoing wave across that boundary. This is discussed in section 5.4.

In general, it is not so easy to find good data to be prescribed on open boundaries. In a sense, these boundary conditions represent the "outside world" and you have only limited knowledge of what is happening outside your model region. Therefore, the information specified through (5.1)...(5.3) is likely to be at least partially wrong. What is the consequence of that? From section 4.4.2 you can conclude that a gravity wave propagating into the interior of the region will be damped at a rate exp($-rt$). In most cases, this is a relatively slow damping, in the sense that the wave will traverse the region without significant decrease in amplitude. This means that the errors in boundary data propagate throughout the area and your solution is no better than the boundary data.

For the "slow" characteristics, the transported quantity is (potential) vorticity, so you should specify

$$q = \frac{\zeta + f}{a} = r(s,t) \quad \text{with} \quad \zeta = \frac{\partial v}{\partial x} - \frac{\partial u}{\partial y} \qquad\qquad (5.4)$$

(only for water particles moving *in*). It is, however, difficult to specify sufficiently detailed information on the potential vorticity. It is difficult to measure due to the gradients involved. You might, of course, assume the incoming vorticity to be zero, but this need not be realistic. Therefore, alternative boundary conditions have been proposed. Oliger & Sundström (1978), for example, argue that it is more natural to specify the tangential velocity component u_s than the potential vorticity q. Another possibility is specifying the flow direction. In all cases, again, your boundary information is likely to be at least partially wrong. From section 4.4.2 you can conclude that a "vorticity wave" is damped due to bottom friction at the same rate as the gravity waves exp($-rt$), so the wrong information will be felt during a time of the order r^{-1}. However, the *distance u/r* travelled during this time is much smaller than for the "fast" waves. You may choose the boundary far enough away from the area of interest for the inaccurate boundary data not to be harmful. The vorticity generated within the model region is then more important than what comes in at the boundary.

So far about purely hyperbolic systems. If you take lateral turbulent exchange into account as an effective viscosity (2.46), the system becomes incompletely parabolic. The "incompleteness" refers to the fact that diffusive terms do not occur in the continuity equation. The theory of characteristics does not apply any longer. Sundström (1977), Oliger & Sundström (1978) applied the energy argument, to be discussed in the next section, to this case. It is obvious anyway that you should provide additional boundary conditions to those discussed before. On a coastline, for example, by analogy to viscous flow you should specify whether it is a *no-slip* boundary, for which the tangential velocity is zero:

$$u_s = 0 \qquad\qquad (5.5)$$

or a *free-slip* boundary, where the shear stress is zero:

$$\partial u_s/\partial n = 0 \qquad\qquad (5.6)$$

On an open boundary, the additional boundary condition is less evident. Sundström proposes

$$\partial u_n/\partial n = 0 \quad \text{on inflow, and}$$
$$\partial u_s/\partial n = 0 \quad \text{on outflow,}$$

the physical significance of which is not very clear.

A general way of providing more complete and coherent boundary data than is usually possible from measurements or assumptions is taking them from a (numerical) model of a larger area. A coarser-grid model of a larger area will produce data on part of the fine-grid boundary points. The remaining ones have to be interpolated. The advantage is that you can use any type of boundary condition that is convenient from the point of view of, e.g., stability (see next section). On the other hand, the difficulty of principle that boundary data may be wrong (i.e. not matching the model behaviour) is not resolved, though the errors will be smaller. The rules for error propagation, given above, still apply. A general question is whether the fine-grid results will be any better than the coarse-grid boundary conditions. A general answer to this question is not possible. If the fine-grid model is dominated by boundary influences, the accuracy is essentially limited by the boundary data. If, on the other hand, phenomena internal to the fine-grid model are dominating, errors in the boundary data are less important.

5.2. Energy arguments

The boundary conditions discussed in the previous section are *necessary* in the sense that the solution is undetermined if you do not provide them. Even if you do, it is not certain that you will get a sensible solution. In certain cases, boundary conditions can lead to instability, i.e. uncontrolled, non-physical growth of the solution (note that this has nothing to do with numerical instability). Therefore, you require the solution to be bounded. A suitable norm to measure growth is energy. The following theory gives conditions for the energy to remain bounded, and, moreover, for the solution to depend continuously on the boundary conditions; together, these properties mean that the mathematical model is well-posed (Kreiss & Oliger, 1973). The conditions are *sufficient*: if you respect them you get a stable solution; if you don't, the theory does not say what happens. Therefore, the theory is only of limited practical value. Yet, the indications on the type of boundary conditions are valuable as guidelines. Moreover, a similar argument may be applied later to some of the numerical methods (chapter 10).

To keep things simple, we discuss only the linearized system (4.3). It turns out to be useful if you make the matrices symmetric by the transformation

$$A' = T A T^{-1}, \quad B' = T B T^{-1}$$

such that A' and B' are symmetric, which can be accomplished by choosing

$$T = \begin{pmatrix} 1 & 0 & 0 \\ 0 & 1 & 0 \\ 0 & 0 & \sqrt{g/a} \end{pmatrix}$$

Multiply (4.3) by T and define $\mathbf{v}' = T \mathbf{v}$, then the system becomes

$$T\frac{\partial \mathbf{v}}{\partial t} + TAT^{-1} T\frac{\partial \mathbf{v}}{\partial x} + TBT^{-1} T\frac{\partial \mathbf{v}}{\partial y} + TCT^{-1} T \mathbf{v} = 0$$

or

$$\frac{\partial \mathbf{v}'}{\partial t} + A'\frac{\partial \mathbf{v}'}{\partial x} + B'\frac{\partial \mathbf{v}'}{\partial y} + C'\mathbf{v}' = 0 \qquad (5.7)$$

with

$$A' = \begin{pmatrix} U & 0 & \sqrt{ga} \\ 0 & U & 0 \\ \sqrt{ga} & 0 & U \end{pmatrix} \qquad B' = \begin{pmatrix} V & 0 & 0 \\ 0 & V & \sqrt{ga} \\ 0 & \sqrt{ga} & V \end{pmatrix} \qquad C' = C$$

To get an energy equation similar to (3.10) (but now for the linearized equations), you premultiply (5.7) by $\mathbf{v}'^{\mathrm{T}}$. Also, take the transpose of (5.7), postmultiply by $\mathbf{v}'$ and add the two results:

$$\mathbf{v}'^{T}\left(\frac{\partial \mathbf{v}'}{\partial t} + A'\frac{\partial \mathbf{v}'}{\partial x} + B'\frac{\partial \mathbf{v}'}{\partial y} + C'\mathbf{v}'\right) + \left(\frac{\partial \mathbf{v}'^{T}}{\partial t} + \frac{\partial \mathbf{v}'^{T}}{\partial x}A'^{T} + \frac{\partial \mathbf{v}'^{T}}{\partial y}B'^{T} + \mathbf{v}'^{T}C'^{T}\right)\mathbf{v}' = 0$$

The system has been symmetrized so that $A'^{T} = A'$; then you get (because the matrices have been assumed constant)

$$\frac{\partial}{\partial t}\mathbf{v}'^{T}\mathbf{v}' + \frac{\partial}{\partial x}\mathbf{v}'^{T}A'\mathbf{v}' + \frac{\partial}{\partial y}\mathbf{v}'^{T}B'\mathbf{v}' + \mathbf{v}'^{T}(C' + C'^{T})\mathbf{v}' = 0 \qquad (5.8)$$

This can then be integrated over the entire region Ω to give a global energy conservation equation for the linearized system, similar to (3.11):

$$\frac{\partial}{\partial t} \| \mathbf{v}'^2 \| + \int_S \mathbf{v}'^T A'_n \mathbf{v}' \, ds + \int \int_\Omega \mathbf{v}'^T (C' + C^T) \mathbf{v}' \, d\Omega = 0 \qquad (5.9)$$

where S is the boundary and

$$\| \mathbf{v}'^2 \| = \int \int_\Omega \mathbf{v}'^T \mathbf{v}' \, d\Omega = \int \int_\Omega \{ u'^2 + v'^2 + \frac{g}{a} h'^2 \} \, d\Omega$$

is the analogue of the energy, expressed in terms of the variations defined in section 4.1. It can be used as a norm of the vector $\mathbf{v}$ as the depth a is positive. In the nonlinear version (3.11), this may be a problem because the depth is part of the solution there and can therefore not be guaranteed to be positive (the physics may include areas that are dry during certain periods of time). See Verboom et al (1982) in this respect. Furthermore, the short-hand notation

$$A'_n = A' \cos \phi + B' \sin \phi$$

has been used where ϕ is the direction of the normal on the boundary.

It is interesting that

$$C' + C^T = \begin{pmatrix} 2r & 0 & 0 \\ 0 & 2r & 0 \\ 0 & 0 & 0 \end{pmatrix}$$

does not contain the Coriolis acceleration f anymore. This agrees with the nonlinear form; the reason is that the Coriolis acceleration does not perform work and therefore should not enter the energy equation.

The core of the energy argument is that (5.9) can be used to see how fast (if at all) the energy may increase. If you can show that it decreases, the solution is bounded. From the same equation you may also prove that the solution then depends continuously on the boundary conditions (this is not shown here). If you consider, first of all, the area integral of bottom friction, you find

$$\int \int_\Omega \mathbf{v}'^T (C' + C^T) \mathbf{v}' \, d\Omega = 2r \int \int_\Omega (u'^2 + v'^2) \, d\Omega \geq 0$$

so bottom friction decreases the energy (as it should). The boundary terms preferably show the same behaviour; if you manage to choose the boundary conditions such that

$$\int_S \mathbf{v}'^T A'_n \mathbf{v}' \, ds \geq 0 \qquad (5.10)$$

you have a (sufficient) condition for stability and well-posedness. If A'_n were positive definite, (5.10) would be true for any vector $\mathbf{v}'$, but this is not the case because A'_n is known to have eigenvalues of both signs. However, it can be written as

$$A'_n = R^T D R$$

where D is a diagonal matrix containing the eigenvalues and R has the corresponding eigenvectors as its rows. These can be found in the same way as in sections 4.3 and 4.4:

$$D = \begin{pmatrix} U_n & 0 & 0 \\ 0 & U_n + \sqrt{ga} & 0 \\ 0 & 0 & U_n - \sqrt{ga} \end{pmatrix} \qquad R = \begin{pmatrix} -\sin\phi & \cos\phi & 0 \\ \dfrac{\cos\phi}{\sqrt{2}} & \dfrac{\sin\phi}{\sqrt{2}} & \dfrac{1}{\sqrt{2}} \\ \dfrac{\cos\phi}{\sqrt{2}} & \dfrac{\sin\phi}{\sqrt{2}} & \dfrac{-1}{\sqrt{2}} \end{pmatrix}$$

Using this, the integrand of (5.10) at any point of the boundary can be rewritten as

$$H = \mathbf{v}^T A'_n \mathbf{v}' = \mathbf{w}^T D \, \mathbf{w} = c_1 w_1^2 + c_2 w_2^2 + c_3 w_3^2 \tag{5.11}$$

where

$$\mathbf{w} = R \; \mathbf{v}' = \begin{pmatrix} u'_s \\ (u'_n + \sqrt{g/a} \; h')/\sqrt{2} \\ (u'_n - \sqrt{g/a} \; h')/\sqrt{2} \end{pmatrix}$$

Please, note the distinction between the mean-flow quantity U_n in D and the velocity variation u'_n, the latter of which is the dependent variable in this linearized analysis. From (5.11) you see that the boundary flux consists of contributions from each of the waves in (4.9); the components of the $\mathbf{w}$ vector are the quantities transported. For the "slow" wave with speed c_1, this is u'_s, which might be a reason to specify this as a boundary condition instead of the potential vorticity.

Firstly, take the case of subcritical outflow. Then $U_n > 0$ is directed outward and you have to specify one boundary condition which prescribes the ingoing component, possibly in terms of the outgoing ones:

$$w_3 = \alpha \, w_1 + \beta \, w_2 + f(t,s) \tag{5.12}$$

with some coefficients α and β. The signal $f(t,s)$ from the outside world states what sort of wave is coming in; it may be disregarded for this analysis. It may be shown that such inhomogeneous terms in the boundary conditions do not change stability. Then

$$H = (c_1 + \alpha^2 c_3) \, w_1^2 + 2\alpha\beta \, c_3 \, w_1 w_2 + (c_2 + \beta^2 c_3) \, w_2^2$$

This should be positive for any w_1 and w_2 as the latter are not known beforehand. This is true only if the expression is positive definite, i.e. if its discriminant is negative and the coefficients of the squares are positive:

$$(\alpha \beta \, c_3)^2 - (c_1 + \alpha^2 \, c_3)(c_2 + \beta^2 \, c_3) \le 0$$
$$c_1 + \alpha^2 \, c_3 \ge 0$$
$$c_2 + \beta^2 \, c_3 \ge 0$$

It turns out that the latter two conditions are included in the first. Substituting the expressions for the propagation speeds, the first condition can be rewritten as

$$\alpha^2 \, (1 - F^2) + \beta^2 \, F \, (1 - F) \le F \, (1 + F) \tag{5.13}$$

where $F = U_n/\sqrt{ga}$ is the Froude number for the normal-flow component; it is between 0 and 1 in this case of subcritical outflow. Eq.(5.13) defines an ellipse in the α, β plane, as illustrated in fig. 5.1. Apparently, you cannot choose arbitrary values for α and β: they must be within the ellipse to get energy decrease. Some important special cases are:

(i) $\alpha = 0$, $\beta = \pm 1$. Then the boundary condition (5.12) reads in the original variables:

$$u'_n - \sqrt{g/a} \; h' = \pm (u'_n + \sqrt{g/a} \; h')$$

so either the normal velocity or the water-level are prescribed (eqs. 5.1 or 5.2). Both choices satisfy (5.13) and therefore do not induce instability.

(ii) $\alpha = 0$, $\beta = 0$ which corresponds to a boundary condition

$$u'_n - \sqrt{g/a} \; h' = 0$$

This is often used as a weakly reflecting boundary condition for the outgoing wave (section 5.4). It obviously satisfies (5.13) and is therefore allowed. Other cases are less interesting but can be checked against the condition of (5.13).

For *subcritical inflow* ($U_n \le 0$ or $-1 < F < 0$), two boundary conditions are needed. You can specify the two ingoing waves in terms of the (single) outgoing one:

$$w_1 = \alpha \, w_2$$
$$w_3 = \beta \, w_2 \tag{5.14}$$

with unknown coefficients α and β (different from the ones above). Again, functions of time can be added to specify a nonzero incoming wave. Then

$$H = (\alpha^2 \, c_1 + c_2 + \beta^2 c_3) \, w_2^2$$

should be positive. Substituting the propagation speeds, this is satisfied if .

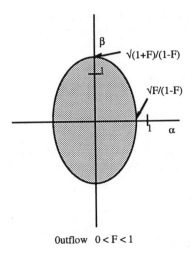

Outflow $0 < F < 1$

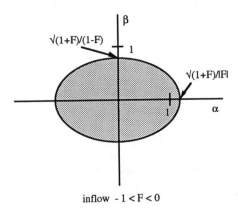

inflow $-1 < F < 0$

Fig. 5.1. Stability regions for outflow and inflow boundaries

$$|F| \, \alpha^2 + (1 - F) \, \beta^2 \leq 1 + F \qquad\qquad (5.15)$$

This represents again an ellipse, shown in fig. 5.1. Special cases are:

(i) $\alpha = 0$, $\beta = \pm 1$ implying boundary conditions

$$u'_n - \sqrt{g/a} \, h' = \pm (u'_n + \sqrt{g/a} \, h')$$
$$u'_s = 0$$

The former corresponds to (5.1) or (5.2). It does *not* satisfy (5.15), i.e. the theory does not prove that the condition leads to a stable formulation. However, as the theory gives only sufficient conditions, this does not really exclude this (very common) type of boundary conditions. Oliger & Sundström (1978), based on a more involved normal-mode theory of Kreiss, suggest that $\beta = -1$ (i.e. specification of normal velocity) is acceptable but $\beta = +1$ (specification of water level) is not on an inflow boundary.

The special case $\alpha = 0$, $\beta = 0$ leads to the same conclusion as above (see further section 5.4).

5.3. Initial conditions

Formally, the plane at time $t = 0$ is also a boundary of the region in x,y,t space. The rule on the number of boundary conditions (here initial conditions) applies just as well. As all characteristics are entering the region across this plane irrespective of the flow direction, you need 3 initial conditions in any case. This means that you will have to specify all three variables h, u, v, which is of course in agreement with what you expect physically.

An arbitrary distribution of these variables over the x-y region at $t = 0$ will break up into waves which will then propagate in their own ways. First of all, you can imagine to decompose the initial data into Fourier components with various wave-number vectors $\mathbf{k} = (k_1, k_2)$. Secondly, each individual component will break up into three contributions corresponding to the three eigenvectors defined in section 4.4. For example, consider the case of pure gravity waves. Assume that you specify as initial conditions:

$$u'(x,y,0) = u_1 \ exp(ik_1x + ik_2y)$$
$$v'(x,y,0) = 0$$
$$h'(x,y,0) = 0$$

with some amplitude u_1. Then the amplitude vector is split up into the three eigenvectors as follows

$$\begin{pmatrix} u_1 \\ 0 \\ 0 \end{pmatrix} = \frac{u_1}{2} \cos \phi \begin{pmatrix} \cos \phi \\ \sin \phi \\ \sqrt{a/g} \end{pmatrix} + \frac{u_1}{2} \cos \phi \begin{pmatrix} \cos \phi \\ \sin \phi \\ -\sqrt{a/g} \end{pmatrix} - u_1 \sin \phi \begin{pmatrix} -\sin \phi \\ \cos \phi \\ 0 \end{pmatrix} \quad (5.16)$$

where ϕ is again the wave direction. The conclusion is that two fast waves with amplitude $u_1/2 \cos \phi$ are generated, propagating in opposite directions, plus a slow wave with amplitude $u_1 \sin \phi$ (and out of phase, indicated by the minus sign). In a similar way you can analyse any other situation.

Again, the problem is that you normally do not know precise initial data, so you have to make assumptions and the initial conditions are inaccurate. The simplest initial condition is that the fluid is at rest with a horizontal water level, but this normally does not happen in reality. Fortunately, the influence of wrong initial data gradually fades out due to wave damping by bottom friction (and, if you have open boundaries that do not fully reflect, also by wave radiation into the "outside world"). In the meantime, the flow gets more and more influenced by external forces or incoming waves, so it "forgets" the initial situation.

As shown in section 4.4, damping works roughly at a rate exp(-*rt*). As an example, *r* may be of the order, say, $0.2 \ 10^{-3}$ s^{-1}, which would mean that the initial wave amplitude is reduced to about 1 % in 6 hours which is indeed the order of magnitude observed in numerical experiments. However, in other cases (small velocities, great depths), *r* can be considerably smaller and the corresponding start-up time correspondingly larger.

How can you determine whether the influence of initial data has been sufficiently reduced? If you have a periodic tide, you can check whether the results for two consecutive tidal periods coincide. Otherwise, a good check might be obtained by starting with two different initial conditions and observing the convergence of the results.

5.4. Reflection

5.4.1. Reflection coefficients

You will not be surprised that a wave hitting a solid boundary (such as a coastline) reflects. Similarly, if you fix the water level at an open boundary, a wave will reflect as well. Actually, any boundary condition gives a certain amount of reflection, so the "open" boundary may not be as open as you would wish. To analyse this, it is useful to determine reflection coefficients for the various kinds of boundary conditions, using the theory of linear harmonic wave propagation developed in chapter 4. To keep things as simple as possible, only one boundary is considered which you can take as $x = 0$ for simplicity. The model region is supposed to be $x \leq 0$ and waves moving in the positive *x*-direction are outgoing.

The boundary condition at $x = 0$ can generally be formulated as

$$\alpha \ u'_n + \beta \ h' = f(y,t) \tag{5.17}$$

Now assume that you have an outgoing wave with amplitude a_1 hitting the boundary; then it will generate a reflected wave with the same frequency ω.

$$\mathbf{v} = a_1 \mathbf{v}_1 exp\{i(k_1 x + k_2 y - \omega t)\} + a_2 \mathbf{v}_2 exp\{i(m_1 x + m_2 y - \omega t)\} \tag{5.18}$$

The amplitude a_2 of the reflected wave is to be determined. The vectors $\mathbf{v}_{1,2}$ are the eigenvectors for the two waves as determined from (4.15). At the boundary $(x = 0)$, the sum of the two waves should satisfy the boundary condition (5.17); this is possible only if the right-hand side has the form

$$f(y,t) = A \ exp\{i(k_2 y - \omega t)\} \tag{5.19}$$

(otherwise a different wave will be generated) and if the wave number component m_2 is equal to k_2. Moreover, both ingoing and outgoing waves satisfy the same dispersion relation (4.16) so $m_1^2 = k_1^2$. The reflected wave then has $m_1 = - k_1$. It is similar to the incident one but the *x*-component of its wave speed is reversed and there is a difference in amplitude (and possibly phase, which is included in the complex number a_2). Up to a normalizing factor, which may be absorbed into the amplitudes, $\mathbf{v}_1$ from (4.15) is found to be

$$\mathbf{v}_1 = \left(- ig \frac{k_1(r - i\omega) + k_2 f}{(r - i\omega)^2 + f^2}, \ - ig \frac{k_2(r - i\omega) - k_1 f}{(r - i\omega)^2 + f^2}, \ 1 \right)^T$$

The eigenvector $\mathbf{v}_2$ for the reflected wave is similar with the sign of k_1 reversed. You can now substitute (5.18) into (5.17); due to our simple choice of coordinates, u_n is just the x component u, so you find

$$a_2 = \frac{\{(r - i\omega)^2 + f^2\}(A - \beta a_1) + ig\alpha \, a_1 \{k_1(r - i\omega) + k_2 f\}}{\beta \{(r - i\omega)^2 + f^2\} - ig\alpha \, \{- k_1(r - i\omega) + k_2 f\}} \tag{5.20}$$

Unless the boundary value A and the outgoing wave a_1 perfectly agree such that the numerator vanishes, there will be an ingoing wave with amplitude a_2. As the outgoing wave is determined by the other boundaries and the interior of the model region, such an agreement is unlikely. The ingoing wave is composed from a part (A) prescribed at the boundary, which could be called the *incoming wave*, and a part generated by the outgoing wave; this is the *reflected wave*. The latter can be expressed in terms of a reflection coefficient R:

$$R = \frac{a_2}{a_1} = - \frac{\beta \{(r - i\omega)^2 + f^2\} - ig\alpha \, \{k_1(r - i\omega) + k_2 f\}}{\beta \{(r - i\omega)^2 + f^2\} - ig\alpha \, \{- k_1(r - i\omega) + k_2 f\}} \tag{5.21}$$

Keep in mind that R is complex; its absolute value gives the amplitude ratio (which is the quantity usually understood as a reflection coefficient) whereas its phase angle gives the phase shift between outgoing and reflected waves. The amount of reflection apparently depends on the direction of wave propagation, given by k_1, k_2.

A few special cases can be recognized. Firstly, if $\alpha = 0$, the reflection coefficient is -1, i.e. you have complete reflection with sign reversal. This is the case when the water level is prescribed. If $f = 0$, the same is true for $\beta = 0$ (normal velocity prescribed, as on a coastline); the reflection coefficient is then +1 without sign reversal. Note that the latter case also occurs if you prescribe the normal velocity at an "open" boundary, which in that case is just as reflecting as a closed one.

Secondly, for pure gravity waves (section 4.4.1), $r = f = 0$ and $\omega = \sqrt{ga}\, k$, so you get

$$R = - \frac{\beta + \alpha \sqrt{g/a} \, \cos \phi}{\beta - \alpha \sqrt{g/a} \, \cos \phi} \tag{5.22}$$

where ϕ is the angle of incidence as before ($\cos \phi = k_1/k$). Now you find complete reflection not only for $\alpha = 0$ (prescribed water level) but also for $\beta = 0$ (prescribed normal velocity, including a coastline).

5.4.2. Weakly reflecting boundary conditions

You could try to define boundary conditions in such a way that you do not get reflection. Eq. (5.20) shows that this is not generally possible; at best it can be obtained for one single wave length and direction. A condition for the outgoing wave has been identified in section 5.2:

$$u'_n - \sqrt{g/a}\ h' = 0 \tag{5.23}$$

This gives $\alpha = 1$ and $\beta = -\sqrt{g}/a$, so from (5.22):

$$R = \frac{\cos\phi - 1}{\cos\phi + 1}$$

which is shown in fig. 5.2 as a function of the angle ϕ. For $\phi = 0$ (normal incidence), this gives $R = 0$: no reflection. This is why (5.23) is sometimes referred to as a nonreflecting boundary condition. However, obviously this is not quite correct, firstly because of the approximations that have been made (specifically in assuming r and f to be zero), but more importantly because there certainly *is* reflection if $\phi \neq 0$; even $R = -1$ if $\phi = \pm\pi/2$, so for waves moving approximately parallel to the boundary, you have almost full reflection.

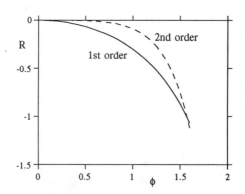

Fig. 5.2. Reflection of pure gravity waves for weakly reflecting boundary conditions: first-order condition (5.23) or second-order condition (5.34)

If you introduce bottom friction ($r \neq 0$), but use the same condition (5.23) as a boundary condition, the only thing that happens is that ω in the reflection coefficient (5.21) should now be determined from (4.17). The reflection coefficient then turns out to be

$$R = -\frac{\beta\,(1 - i\omega/r)\sqrt{r^2/ga\,k^2} - i\,\alpha\sqrt{g/a}\,\cos\phi}{\beta\,(1 - i\omega/r)\sqrt{r^2/ga\,k^2} + i\,\alpha\sqrt{g/a}\,\cos\phi} \tag{5.24}$$

With $\alpha = 1$ and $\beta = -\sqrt{g}/a$ this is nonzero even for normal incidence, as shown in fig. 5.3a. Depending on the magnitude of the bottom friction coefficient, 20 to 30 % reflection may occur, which is quite appreciable, though much better than full reflection obtained from a "hard" condition like (5.1) or (5.2). A phase shift dependent on the angle of incidence is also introduced: see fig. 5.3b

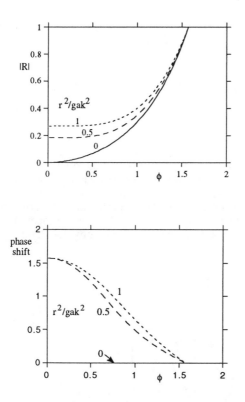

Fig. 5.3. Reflection coefficient for weakly-reflecting boundary condition (5.23) with bottom friction included. Top: absolute value; bottom: phase shift.

Many attempts have been made to improve on this reflective behaviour of the boundary conditions. The way to do that has been indicated by Engquist & Majda (1977), see also Verboom & Slob (1984), Higdon (1986, 1987). The idea is relatively simple; unfortunately the elaboration generally is not, so only an example is shown. Consider as before a boundary at $x = 0$. The set of equations (4.3) is Fourier transformed in t and y directions; the transformed variables are called ω and k_2, respectively. They correspond to variables with the same names used above. We omit in this example the lower-order terms $(C = 0)$. This gives

$$A \frac{\partial \overline{v}}{\partial x} + (ik_2 B - i\,\omega)\,\overline{v} = 0 \tag{5.25}$$

$\overline{v}$ indicating the Fourier-transformed variables. Similar to the wave-propagation analysis, you can express the vector $\overline{v}$ in terms of a set of eigenvectors:

$$\overline{\mathbf{v}} = \sum_{j=1}^{3} w_j(x)\, \mathbf{e}_j \tag{5.26}$$

where $\mathbf{e}_j$ indicates an eigenvector with corresponding eigenvalue λ_j according to

$$(\lambda_j A + i k_2 B - i\omega I)\, \mathbf{e}_j = 0 \tag{5.27}$$

Introducing this into (5.25) splits the system into three scalar components:

$$\frac{\partial w_j}{\partial x} + \lambda_j\, w_j = 0 \tag{5.28}$$

If any λ_j has a negative imaginary part, this corresponds to an ingoing wave; for such a wave you should specify a boundary condition, i.e. you should prescribe the value of the corresponding w_j at the boundary. To see what this means, suppose $V = 0$. You cannot take $U = 0$ as well because you have seen that the number of boundary conditions depends on the sign of U. Let us take $U > 0$ (outflow). Then you expect that one boundary condition is needed. It is possible to figure out the eigenvalues and eigenvectors from (5.27) in this case; the result is

$$\lambda_1 = \frac{i\omega}{U} \qquad \lambda_{2,3} \approx \pm i\ \sqrt{\frac{\omega^2}{ga} - k_2^2} \tag{5.29}$$

$$\mathbf{e}_1 = \begin{pmatrix} -k_2 U \\ \omega \\ 0 \end{pmatrix} \qquad \mathbf{e}_{2,3} \approx \begin{pmatrix} \pm g\sqrt{\omega^2/ga - k_2^2} \\ g\ k_2 \\ \omega \end{pmatrix} \tag{5.30}$$

In the components labeled 2,3, U has been taken zero in the result, which implies that these expressions are valid only for small Froude number F. As it is likely that $\omega^2 > ga\, k_2^2$ the eigenvalues are all imaginary; two are positive and correspond to outgoing waves; one is negative indicating an ingoing wave; assume that this is λ_3. Then obviously, you should specify w_3 as a boundary condition. The other components are outgoing and should be left unspecified. Using (5.30), you can find from (5.26) that

$$w_3 \approx \frac{\overline{h}}{2\omega} - \frac{\overline{u}}{2g\ \sqrt{\omega^2/ga - k_2^2}}$$

where again terms of the order U have been neglected. A homogeneous boundary condition would be to set this quantity to zero:

$$g\ \sqrt{\omega^2/ga - k_2^2}\ \overline{h} - \omega\overline{u} = 0 \tag{5.31}$$

However, to be useful, it should be transformed back to physical space and that is where the difficulties come in. A factor ω corresponds to differentiation with respect to time, but ω occurs in a complex manner in (5.31); similarly for k_2. Formally Fourier-transforming (5.31) back to physical space results in a convolution integral over the boundary and over time, i.e. in a nonlocal and non-instantaneous boundary condition. This would be very awkward to handle. Therefore, you could try to approximate (5.31), e.g. in this way, assuming k_2 to be small (large wave length):

$$\sqrt{\frac{g}{a}}\, \omega \left(1 - \frac{g a k_2^2}{2\omega^2}\right) \bar{h} - \omega \bar{u} = 0 \tag{5.32}$$

Neglecting the term with k_2 and dividing by ω, this gives

$$\sqrt{\frac{g}{a}}\, \bar{h} - \bar{u} = 0$$

which can be immediately transformed resulting in (5.23). The latter can therefore be considered a first-order approximation of a non-reflecting boundary condition. Curiously, you need not divide by ω, then you find after inverse transformation

$$\frac{\partial}{\partial t}\left(\sqrt{\frac{g}{a}}\, h - u\right) = 0$$

which is almost but not exactly equivalent to (5.23) Apparently, there is a certain freedom of choice. A more accurate condition is obtained if you do not neglect the term with k_2 in (5.32). Rewriting it gives

$$\omega^2 \left(\bar{u} - \sqrt{\frac{g}{a}}\, \bar{h}\right) + \frac{1}{2}\sqrt{\frac{g}{a}}\, g a\, k_2^2\, \bar{h} = 0$$

which transforms into what can be called a second-order approximation to a non-reflecting boundary condition (this is again not unique):

$$\frac{\partial^2}{\partial t^2}\left(u - \sqrt{\frac{g}{a}}\, h\right) + \frac{1}{2}\sqrt{\frac{g}{a}}\, g a\, \frac{\partial^2 h}{\partial y^2} = 0 \tag{5.33}$$

What is the reflection coefficient for this condition? For a harmonic wave, (5.33) can be written in the form (5.17) with

$$\alpha = 1, \qquad \beta = -\sqrt{\frac{g}{a}}\left(1 - \frac{g a k_2^2}{2\omega^2}\right)$$

Substituting this into (5.22) gives after some rearrangement

$$R = \frac{\cos\phi - 1 + \frac{1}{2}\sin^2\phi}{\cos\phi + 1 - \frac{1}{2}\sin^2\phi} \tag{5.34}$$

which is illustrated in fig. 5.2. You see that there is much less reflection than with the first-order condition, particularly for small angles of incidence. For large angles, there is no advantage of the more complicated condition.

In principle, the analysis can be extended with full advection terms, bottom friction or Coriolis acceleration, but the analysis is very tedious and the resulting boundary conditions are complicated. Some proposals have been put forward by Engquist & Majda (1977) and Verboom & Slob (1984). Very few actual applications of such higher order boundary conditions are known; usually, one is satisfied with (5.23). Verboom & Slob give an example. A warning is in order that not all boundary conditions derived in this way lead to a stable model formulation. Other possible open-boundary conditions have been discussed by Røed & Cooper (1987).

5.4.3. Specification of an incoming wave

Eq. (5.20) gives the recipe how to specify an incoming wave. As an example, suppose you want to have a wave moving *into* the region, having a frequency ω, wave-crest direction ϕ and amplitude of water-level variation H. The wave length will be selected through the dispersion relation (4.16): you insert $k_1 = k \cos \phi$, $k_2 = k \sin \phi$ and solve for k. If the water level amplitude in the eigenvector is normalized to 1, (5.18) shows that you should set a_2 to the required amplitude H. Finally, the non-homogeneous part of (5.20) shows that you should choose

$$A = \frac{\beta \{(r - i\omega)^2 + f^2\} - ig\alpha \{- k_1(r - i\omega) + k_2f\}}{\{(r - i\omega)^2 + f^2\}}H \qquad (5.35)$$

in which you can still insert the values of α and β you want to use. You can combine various terms of this kind in the right-hand side of (5.17) to get a combination of incoming waves with different properties. Even more generally, you could specify any function $f(y,t)$ which will then be "interpreted" by the model as a superposition of waves with various frequencies, directions and amplitudes.

5.5. Moving boundaries

In shallow coastal areas, parts of the region may dry up at low water levels, which means that you have moving boundaries. The definition of these boundaries would appear to be rather straightforward: the water depth $a = 0$. However, there are some difficulties involved. First, with this condition, the propagation speeds all coalesce into a single one, so that the hyperbolic character of the equations gets lost and it is not so obvious that you are left with a mathematically well-posed problem. Second, the bottom friction terms in (2.51) and (2.52) have a singular character, so you would probably be well advised to use the conservative form of (2.48),(2.49) instead; but there, too, it is not obvious what happens if $a = 0$. The whole question has not been very well resolved. Numerically, a number of more or less ad-hoc methods have been used, usually with a lot of difficulties to prevent numerical instability: examples are Sielecki & Wurtele (1970), Johns et al (1982), Stelling (1983), Gopalakrishnan (1989).

Chapter 6

Discretization in space

6.1. Finite differences

In this chapter, the spatial discretization of the shallow-water equations is discussed. The most common approach is that of finite differences, in which the region of interest is covered by a rectangular grid with grid sizes Δx, Δy and the spatial derivatives are approximated by finite differences. For example, a central x-difference is defined as

$$D_{0x}u \equiv \frac{u_{k+1,j} - u_{k-1,j}}{2\Delta x} \approx \frac{\partial u}{\partial x} \tag{6.1}$$

where u_{kj} indicates the value of the variable u at grid point $(k\Delta x, j\Delta y)$. A forward difference is defined as

$$D_{+x}u \equiv \frac{u_{k+1,j} - u_{k,j}}{\Delta x} \approx \frac{\partial u}{\partial x} \tag{6.2}$$

and a backward difference D_{-x} does the same thing in the opposite direction. These are less accurate than a central difference. Similar operations in y-direction are indicated by a subscript y. Using central differences, the linearized SWE (4.3) are approximated by

$$\frac{\partial \mathbf{v}}{\partial t} + A\, D_{0x}\mathbf{v} + B\, D_{0y}\mathbf{v} + C\mathbf{v} = 0 \tag{6.3}$$

Time is kept continuous (time discretization is discussed in ch. 8), so you get a *semi-discrete* system of ordinary differential equations for the values h, u, v in all grid points as functions of time. At boundaries, central differences cannot be used as you would need grid points outside the region. Special procedures are needed, which are discussed in chapter 10.

A measure of accuracy is the truncation error, which is obtained by substituting the exact solution into the finite-difference equation (6.3). Of course, the exact solution is usually unknown but it can be done as a theoretical exercise. Expanding all terms into Taylor series relative to the central point k,j; and showing only the few lowest order terms you find

$$\frac{\partial \mathbf{v}}{\partial t} + A\frac{\partial \mathbf{v}}{\partial x} + B\frac{\partial \mathbf{v}}{\partial y} + C\mathbf{v} = \frac{1}{6}\Delta x^2 A\frac{\partial^3 \mathbf{v}}{\partial x^3} + \frac{1}{6}\Delta y^2 B\frac{\partial^3 \mathbf{v}}{\partial y^3} + \dots \tag{6.4}$$

where the dots represent higher-order terms, supposed to be less important. You could say that (6.4) is the equation you are actually solving; it differs from the desired equation by the right-hand side terms which constitute the *truncation error*. Eq. (6.4) is sometimes called the *modified equation*. Note that the truncation error is the error in the *equation*, not in the solution. It is of second order in Δx and Δy, which means that the error will be reduced by a factor of 4 if you halve the grid sizes. Unfortunately, this does not say very much in a quantitative sense. For that purpose the wave-propagation analysis of chapter 7 is more useful.

In actual practice, you want to discretise the full nonlinear SWE. You have the choice between the conservation-law form of (2.48-50) and the differential form (2.51-52). The latter case can be written in the same form as above with the understanding that *A*, *B* and *C* are now variable; they should then be evaluated at the central point *k*, *j*. In the conservative form, differences of fluxes are taken, e.g. for (2.48):

$$\frac{\partial h}{\partial t} + \frac{(au)_{k+1,j} - (au)_{k-1,j}}{2\Delta x} + \frac{(av)_{k,j+1} - (av)_{k,j-1}}{2\Delta y} = 0 \qquad (6.5)$$

and similarly for the other equations. As noted in section 3.4, you have to use the conservative form if you want to represent discontinuous solutions correctly. For smooth solutions (long waves), there are no strong arguments to favour one formulation over the other. The *order* of approximation is the same for both but there may be a quantitative difference in accuracy, amongst others in terms of mass and momentum conservation (section 7.8). In a linearized wave propagation analysis, they are equivalent.

6.2. Staggered grids

If you simplify the equations even further by omitting advection ($U = V = 0$) and Coriolis terms ($f = 0$), and trace which variables are needed in each equation, you get the scheme shown in fig. 6.1. The interesting thing is that you do not need all the variables at all points: the subset shown in the figure is sufficient to solve the equations; it is called a *staggered* grid. Actually, there are four such grids, the other three being obtained by shifting the one in fig. 6.1 over one grid interval north, west or northwest. By superposing these four subgrids, you get a full grid with all variables defined at each grid point (unstaggered grid).

One consequence of this is that you can cut your computation cost by a factor of 4 by using just one of the subgrids, without sacrificing any accuracy (at least in principle; below you will see that things are not quite that simple). Secondly, if you would decide to use all four subgrids, the four partial solutions do not have anything in common except perhaps at the boundaries. This means that the solutions are decoupled, or only weakly coupled. Numerically, the solutions may grow apart. Looking at the full grid, this gives the impression of a short-wave oscillation, like a checkerboard, as nearby grid values for the same variable belong to different subgrids and therefore appear to be tending towards (slightly) different values. This makes the full grid less attractive.

The staggered grid of fig. 6.1. is one of several possibilities, known as the Arakawa A ... E grids (fig. 6.2, Arakawa 1972). These grids are shown in such a way that they have the same grid size relative to wave length when applied to the simplified SWE as discussed in this section. Grid A is fully unstaggered; grid C is identical with fig. 6.1. Grid E is one of

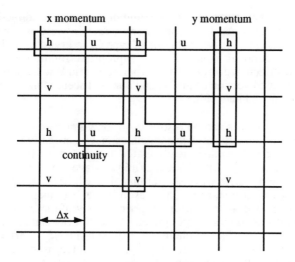

Fig. 6.1. Grid values needed in the simplified finite-difference equations

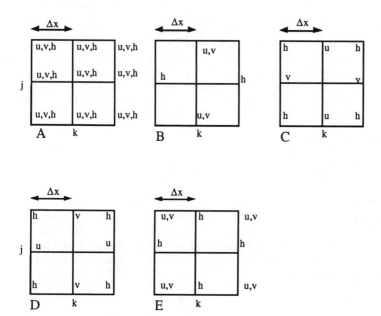

Fig. 6.2. Grid layout for Arakawa A...E grids

two subgrids, the other one being obtained by shifting in north or west direction; the two together give one "full" grid A. Therefore, here you win only a factor of 2 in computational cost. Grid E contains two subgrids of type C, so grid decoupling may still occur. Grid B is almost identical to grid E with appropriate rotation, redefinition of velocity components and effective grid size $\Delta x \sqrt{2}$. Therefore, grid B will not be considered separately. Grid D is a rotated version of grid C and does not seem to have any particular advantage, so it is not discussed here either.

The grids differ in the way waves can be resolved. In one dimension, it is known that waves with less than two grid points per wave length cannot be properly represented. If k is the wave number and $\xi = k\Delta x$, this means that $-\pi \le \xi \le \pi$ (or if you allow only positive wave numbers, $0 \le \xi \le \pi$). The maximum wave number $k_{max} = \pi/\Delta x$ corresponds to a wave length $2\Delta x$. A wave shorter than this will be interpreted as a longer one; the phenomenon is called *aliasing*. The reason is that a sequence of grid values $u_j = \sin j\xi$ is identical to $\sin (j\xi + 2\pi m)$ with integer m. If any of these corresponds to a wave length $> 2\Delta x$, that is how it will be interpreted. We will show that this is possible if $\xi > \pi$. Take, for example, $\pi < \xi < 3\pi$ and choose $m = -j$, then $\sin (j\xi - 2\pi j) = \sin j\xi'$ with $\xi' = \xi - 2\pi$ which is in the resolvable interval: ξ is interpreted as $\xi - 2\pi$, or k as $k - 2k_{max}$. Similarly, if $3\pi < \xi < 5\pi$, the apparent wave has $\xi' = \xi - 4\pi$ etc.

In the 2-d case, a similar phenomenon occurs: a field of grid values $u_{i,j} = \sin (i\xi + j\eta)$ with $\xi = k_1\Delta x$ and $\eta = k_2\Delta y$ (i is just an integer here, not the imaginary unit) may be interpreted as $\sin (i\xi' + j\eta')$ with $\xi' = \xi + 2\pi n$ and $\eta' = \eta + 2\pi m$ for some integers m,n. That means that the resolvable range of wave numbers is as shown in fig. 6.3. for grid A. The same reasoning applies to each variable u, v, h separately. Grids B, C and D all have grid points for a single variable at double grid size. The resolvable wave-number range is therefore reduced by a factor 2. This is not necessarily a disadvantage, as waves in the region $\pi/2 < \xi,\eta \le \pi$ are poorly represented, subject to important numerical errors (section 7.1) and sometimes the cause of nonlinear instabilities. So, actually the reduced range of resolved wave numbers is often mentioned as an argument *in favor* of staggered grids. For grid E (assuming equal grid sizes in both directions), grid values can be described as

$$u = \sin (k_1 x + k_2 y) = \sin (k'_1 x' + k'_2 y')$$

in a grid with grid size $\Delta x \sqrt{2}$, rotated over an angle $\pi/4$:

$$x' = \frac{1}{2} (x + y) \qquad y' = \frac{1}{2} (x - y)$$
$$k'_1 = k_1 + k_2 \qquad k'_2 = k_1 - k_2$$

where k_1, k_2 are the original wave-number components. In the rotated grid, the resolvable wave numbers should satisfy $-\pi \le \xi'$, $\eta' \le \pi$. Translating this back to the original coordinates, you find the resolvable wave number region as shown in fig. 6.3:

$$-\frac{\pi}{\sqrt{2}} \le \xi + \eta \le \frac{\pi}{\sqrt{2}} \quad \text{and} \quad -\frac{\pi}{\sqrt{2}} \le \xi - \eta \le \frac{\pi}{\sqrt{2}}$$

The discretization on staggered grids is a little more complicated if you reintroduce the neglected terms. If grid values are not available where you need them for certain terms, you will either have to use grid points at greater distance or average between adjacent grid points.

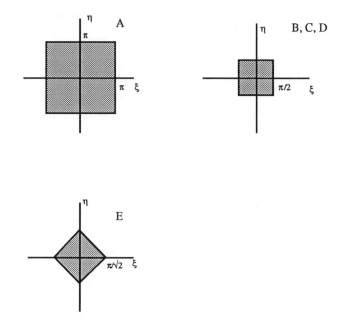

Fig. 6.3. Resolvable wave-number regions for grids A ... E

For example, take grid C and consider the equation of continuity at grid point k,j: you would need, e.g., *(au)* at point $k+1,j$ but a (or h) is not defined there, so you would have to average between points k,j and $k+2,j$. This has a consequence for the accuracy. Moreover, the number of operations increases due to averaging, so you do not gain quite the factor of 4 normally expected for grid C.

For the linearized SWE, the semi-discrete equations on grid E can no longer be written in a nice simple matrix form such as (6.3) as various terms have to be treated in different ways. In h-points and u,v-points, separate equations are valid:

$$\frac{\partial h}{\partial t} + U M_y D_{0x}h + V M_x D_{0y}h + a\{D_{0x}u + D_{0y}v\} = 0$$

$$\frac{\partial u}{\partial t} + U M_y D_{0x}u + V M_x D_{0y}u - fv + gD_{0x}h + ru = 0 \qquad (6.6)$$

$$\frac{\partial v}{\partial t} + U M_y D_{0x}v + V M_x D_{0y}v + fu + gD_{0y}h + rv = 0$$

where averaging operators $M_{x,y}$ have been introduced; expressed symbolically in terms of the weights of 3*3 grid points around k,j:

$$M_x = \frac{1}{2}\begin{bmatrix} 0 & 0 & 0 \\ 1 & 0 & 1 \\ 0 & 0 & 0 \end{bmatrix}$$

and similarly for M_y. Consequently,

$$M_y D_{0x} = \frac{1}{4\Delta x} \begin{bmatrix} -1 & 0 & 1 \\ 0 & 0 & 0 \\ -1 & 0 & 1 \end{bmatrix}$$

The averaging operators introduce additional terms in the truncation error, for example

$$M_y D_{0x} h_{k,j} = \tfrac{1}{2}(D_{0x} h_{k,j+1} + D_{0x} h_{k,j-1}) = \frac{\partial h}{\partial x} + \tfrac{1}{6}\Delta x^2 \frac{\partial^3 h}{\partial x^3} + \tfrac{1}{2}\Delta y^2 \frac{\partial^3 h}{\partial x \partial y^2} + \dots$$

the latter term of which is due to the averaging. You see that the *order* of the truncation error does not change, but there may be a quantitative effect on the accuracy (see further section 7.1). The nonlinear (advective) form to be used in actual practice looks like

$$\frac{\partial h_{k,j}}{\partial t} + \frac{u_{k+1,j} M_y a_{k+1,j} - u_{k-1,j} M_y a_{k-1,j}}{2\,\Delta x} + \frac{v_{k,j+1} M_x a_{k,j+1} - v_{k,j-1} M_x a_{k,j-1}}{2\,\Delta y} = 0$$

$$\frac{\partial u_{k,j}}{\partial t} + u_{k,j} M_y \frac{u_{k+1,j} - u_{k-1,j}}{2\,\Delta x} + v_{k,j} M_x \frac{u_{k,j+1} - u_{k,j-1}}{2\,\Delta y} - f v_{k,j} + g D_{0x} h_{k,j} +$$

$$+ c_f \frac{u_{k,j}}{M_1 a_{k,j}} \sqrt{u_{k,j}^2 + v_{k,j}^2} = 0$$

(6.7)

and similarly for the *y*-direction. Note that the *k,j* values in the two equations are different. In the bottom-friction term, there is no good reason to average the depth in one direction only, so a 4-point average has been used, the stencil of which is

$$M_1 = \frac{1}{4} \begin{bmatrix} 0 & 1 & 0 \\ 1 & 0 & 1 \\ 0 & 1 & 0 \end{bmatrix}$$

The truncation error due to this operation is again of second-order.

For grid C, a somewhat different way of averaging is needed. Advective terms must be averaged in their own direction as grid values in the other direction are not available. Moreover, averaging is now also needed in the Coriolis terms as the velocity components are defined in different grid points.

$$\frac{\partial h}{\partial t} + U M_x D_{0x} h + V M_y D_{0y} h + a\{D_{0x} u + D_{0y} v\} = 0$$

$$\frac{\partial u}{\partial t} + U M_x D_{0x} u + V M_y D_{0y} u - f M_x M_y v + g D_{0x} h + ru = 0$$

(6.8)

$$\frac{\partial v}{\partial t} + U M_x D_{0x} v + V M_y D_{0y} v + f M_x M_y u + g D_{0y} h + rv = 0$$

Note that $M_x D_{0x}$ is actually a central difference with double grid size, e.g.

$$M_x D_{0x} h_{k,j} = \frac{h_{k+2,j} - h_{k-2,j}}{4\Delta x} = \frac{\partial h}{\partial x} + \frac{2}{3}\Delta x^2 \frac{\partial^3 h}{\partial x^3} + \ldots$$

so again the order of approximation remains unchanged but the error is four times larger than on an unstaggered grid. The combination $M_x M_y$ is an oblique 4-point average:

$$M_x M_y = \frac{1}{4}\begin{bmatrix} 1 & 0 & 1 \\ 0 & 0 & 0 \\ 1 & 0 & 1 \end{bmatrix}$$

Alternatively, you could average the Coriolis terms in one direction only, but this would involve a double grid size again and the error would be larger. The nonlinear equations can be discretized on grid C as

$$\frac{\partial h_{k,j}}{\partial t} + \frac{u_{k+1,j} M_x a_{k+1,j} - u_{k-1,j} M_x a_{k-1,j}}{2\,\Delta x} + \frac{v_{k,j+1} M_y a_{k,j+1} - v_{k,j-1} M_y a_{k,j-1}}{2\,\Delta y} = 0$$

(6.9)

$$\frac{\partial u_{k,j}}{\partial t} + u_{k,j} M_x D_{0x} u_{k,j} + M_x M_y v_{k,j} M_y D_{0y} u_{k,j} - f M_x M_y v_{k,j} + g D_{0x} h_{k,j} +$$

$$+ c_f \frac{u_{k,j}}{M_x a_{k,j}} \sqrt{u_{k,j}^2 + (M_x M_y v_{k,j})^2} = 0$$

and similarly for the y-direction. You will notice that in the nonlinear versions essentially the same averaging has to be done as in the linear ones, which is one of the reasons to study accuracy for the linear equations only (the other reason being that we do not have simple means to study solutions of nonlinear equations). As far as the truncation error is concerned, you will find additional error terms due to averaging, all of which are of second order. Nevertheless, in the nonlinear equations, some more averaging is involved in the advective and bottom-friction terms. The way of averaging will be obvious if you look at the available values in the grid stencil.

6.3. Curvilinear grids

In actual practice, coastlines are not usually rectangles, so you get into trouble when using a simple uniform rectangular or square grid (fig. 6.4a) as the boundary will not coincide with grid points. A traditional way out is to use an approximate boundary formed by the nearest grid points as shown in the same figure. Points outside the region are labelled "inactive" and do not participate in the computation, although they usually are present in computer memory. Apart from the inefficiency involved, there are problems with numerical accuracy as well .

Therefore you may try to use more flexible grids which are aligned with the boundaries and perhaps have a varying grid size to have greatest accuracy where the strongest gradients are expected. A simple but not entirely satisfactory way to do this is a "telescoping grid" in which rectangular grid lines are pushed in or out such that grid points are forced to be on the boundary (fig. 6.4.b). The finite-difference formulas must be

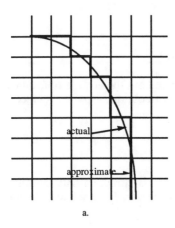 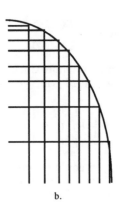

a.   b.

Fig. 6.4. Approximation of curved boundary
(a) on a rectangular grid and (b) on a telescoping grid.

adjusted to account for uneven grid sizes surrounding a particular grid point. The figure
shows that this method may give quite strange grids away from the boundaries. Therefore,
this approach has never become very popular.

A third approach for finite-difference methods, *curvilinear boundary-fitted grids* has
become quite popular during the last 15 years (see , e.g. Johnson, 1982 or Willemse et al,
1985, for application to the SWE). Structured grids are used, which means that although a
region is not a rectangle, the arrangement of grid points is in rectangular form: each row
of grid points contains the same number of points even though they are not on straight
lines (fig. 6.5). This has great advantages for programming as the difference equations
will have the same structure as in the rectangular grid. The number of neighbouring points
is known and they can be referenced by subscripts k,j in just the same way.

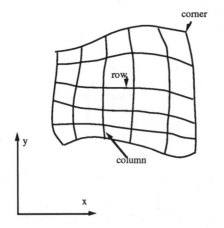 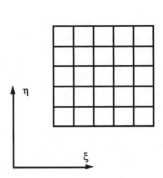

Fig. 6.5. Curvilinear grid in physical and computational domains

A formal coordinate transformation is introduced from the physical region into a computational region which is rectangular (fig. 6.5) as described in section 2.10. All computations are done in the computational region where you have a simple rectangular grid. The price to be payed is that the equations are more complicated.

You have the choice to work with either Cartesian velocity components (2.54 or 2.55) or contravariant components (2.63 ... 2.65 or 2.67). In the former case, the velocity components need not be aligned with the grid. In chapter 7, you will see the effect of this in terms of numerical accuracy. The advantage is that the equations remain relatively simple. The approach was used by Johnson (1982), Borthwick & Barber (1992), the latter with a non-orthogonal grid. The contravariant form utilizes the velocity components related to the grid. The main additional complication is the occurrence of Christoffel symbols containing second derivatives of the coordinate transformation, which are numerically awkward. For the orthogonal case, this approach was used by Willemse et al (1985).

Discretization of the equations in curvilinear form is usually done by the finite-volume method, either in the physical region or *in the computational region*; the two ways are very closely related.

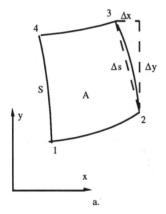   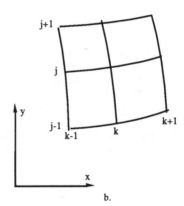

Fig. 6.6. Curvilinear grid

The finite-volume method is based on the conservative form of the equations, which are integrated over the finite volumes (or surfaces in a 2-d sense) defined by the grid lines. If you take, for example, the conservation equation for mass (2.48)

$$\frac{\partial h}{\partial t} + \frac{\partial f}{\partial x} + \frac{\partial g}{\partial y} = 0$$

(with $f = au$, $g = av$ the mass fluxes), integration over a cell in the physical (x,y) region will give (see fig. 6.6a), using Gauss' theorem

$$\frac{\partial}{\partial t} \int \int_A h \, dxdy + \sum_{j=1}^{4} \int_{S_j} f_n \, ds = 0 \qquad (6.10)$$

with $f_n = (f, g).\mathbf{n}$ denoting the flux component normal to the cell face. The cell average can be expressed in the four grid values by numerical integration. The fluxes across the cell faces can be computed relatively easily if the faces are straight lines:

$$f_n = f \frac{\Delta y}{\Delta s} - g \frac{\Delta x}{\Delta s}$$

so by the trapezoidal rule (A is the surface area of the cell)

$$2A \frac{\partial h}{\partial t} + (f_1 + f_2)(y_2 - y_1) - (g_1 + g_2)(x_2 - x_1) + (f_2 + f_3)(y_3 - y_2) - (g_2 + g_3)(x_3 - x_2) +$$
$$+ (f_3 + f_4)(y_4 - y_3) - (g_3 + g_4)(x_4 - x_3) + (f_4 + f_1)(y_1 - y_4) - (g_4 + g_1)(x_1 - x_4) = 0$$
$$(6.11)$$

which can be used even if the grid is irregular. The accuracy of this discretisation is not easy to determine by means of Taylor series expansions. However, you can put in polynomials of x and y and the highest order for which the formula gives the exact result will be the order of the truncation error. Now, the $\partial f / \partial x$ part of (6.11) can be written as

$$(f_3 - f_1)(y_4 - y_2) + (f_2 - f_4)(y_3 - y_1)$$

If f = constant, this gives zero, which is the exact result. If $f = x$, you find

$$(x_3 - x_1)(y_4 - y_2) + (x_2 - x_4)(y_3 - y_1)$$

and this happens to be $2A$ (as you may check from two connected triangles), so after division by $2A$, you get the correct result $\partial f / \partial x = 1$. Similarly, if $f = y$, you get zero. However, if $f = y^2$, the result is

$$(y_3 + y_1 - y_4 - y_2)(y_3 - y_1)(y_4 - y_2)$$

and this is generally not zero as it should. The conclusion is that only linear polynomials are handled exactly and the method has a first order truncation error for irregular grids. The other equations can be handled the same way with a similar result. Due to the limited accuracy obtained, this approach is not generally recommended.

On the other hand, the finite-volume approach can also be applied in the computational region. Eq. (6.10) then looks like

$$\int \int_A \frac{\partial h}{\partial t} J \, d\xi d\eta + \int_{\substack{2-3 \\ 4-1}} (y_\eta f - x_\eta g) \, d\eta + \int_{\substack{1-2 \\ 3-4}} (-y_\xi f + x_\xi g) \, d\xi = 0 \qquad (6.12)$$

If the transformation is smooth, an alternative can be obtained by extending the integration over four cells (fig. 6.6b). The surface integral can then be approximated by a single-point integration and the fluxes by the mid-point rule:

$$A_{kj} \frac{\partial h_{kj}}{\partial t} + 2 \, \Delta\eta \left[y_\eta f - x_\eta g \right]_{k-1,j}^{k+1,j} + 2 \, \Delta\xi \left[-y_\xi f + x_\xi g \right]_{k,j-1}^{k,j+1} = 0 \qquad (6.13)$$

where A is now the area of the four cells. This discretisation reverts to standard central differencing if the grid is rectangular and uniform. Eq.(6.13) is a second-order approximation in the computational region if the functions to be integrated are smooth, *including the transformation coefficients* (see also Fletcher, 1988). More precisely, f, g, x_η and y_η should be twice differentiable in ξ direction and x_ξ and y_ξ in η direction. Together, this means that x and y should be twice differentiable in both directions. If this condition is not met, the order of approximation drops to first or even zeroth order. This is a convergence question. For a grid of a certain finite size, it is very difficult to define exactly what the degree of smoothness should be and whether the accuracy of the result is influenced by it. You just have to be aware that there could be such an influence. At least, grids with sudden jumps or kinks in the grid distribution should be viewed with suspicion. One of the ways to generate a grid is solving Poisson's equations for x and y (see section 6.6). This produces grids with the required smoothness, provided the coefficients in these equations are continuous.

The variables involved in (6.13) are still the original Cartesian velocity components. You may also use the formulation in terms of contravariant velocity components (2.62, 2.67). Integration over an area of 2*2 grid cells in the ξ,η plane with single-point integration on the cell sides yields for the mass-balance equation (2.62):

$$4 \, \Delta\xi \, \Delta\eta \frac{\partial h_{kj}}{\partial t} + 2 \, \Delta\eta \left[a \, v^1 \right]_{k-1,j}^{k+1,j} + 2 \, \Delta\xi \left[a \, v^2 \right]_{k,j-1}^{k,j+1} = 0 \qquad (6.14)$$

This is equivalent with (6.13) if you replace the velocity components using (2.59) and the transformation coefficients using (2.57). Finally, you use the fact that

$$4 \, \Delta\xi \, \Delta\eta J = A_{kj}$$

For the momentum equations, things are somewhat more complicated due to the occurrence of second-order tensors. Integrating (2.67) over 4 grid cells gives the contribution

$$\int\int_A (a v^i v^j)_{,j} d\xi \, d\eta = \int\int_A \left(\frac{\partial}{\partial \xi^j}(a v^i v^j) + \left\{ \begin{matrix} i \\ p \ k \end{matrix} \right\} a v^p v^j + \left\{ \begin{matrix} j \\ p \ k \end{matrix} \right\} a v^i v^p \right) d\xi \, d\eta$$

$$(6.15)$$

The first part of this can be evaluated in a straightforward way similar to (6.14); the second and third terms must be approximated by

$$4\,\Delta\xi\,\Delta\eta\!\left(\left\{ \begin{array}{cc} i \\ p\ k \end{array}\right\} av^p v^j + \left\{ \begin{array}{cc} j \\ p\ k \end{array}\right\} av^i v^p \right)$$

at the central point. This is a second order approximation if the metric coefficients (and, as a matter of fact, the Christoffel symbols) are smooth. The pressure term gives

$$\int\int_A g^{ij}\frac{\partial h}{\partial\xi^j}\,d\xi\,d\eta$$

which must be approximated in the center point and using finite differences for the gradients. This requires again smoothness of the metric coefficients. In this way, you can arrive at similar finite-difference formulations as in the case of a rectangular grid, but with some terms added. It is also possible to define a staggered grid in the computational plane. The finite-volume integrations should then be taken on a cell with faces halfway the grid points. The cell fluxes can then be approximated using the available surrounding grid values. As there is insufficient experience available to judge whether any approach might be better than the other, further details are not given

6.4. Finite elements

The finite-element method (FEM) has not been very popular in the field of the SWE, even though it is very flexible for the representation of complicated geometries. The reason is probably that finite-element methods, if used with their traditional implicit time discretisation (see chapter 8), tend to be very expensive. However, there is no need to use an implicit form. Actually, the FEM is a method of spatial discretisation and you can choose any method for integrating in time. A description of the FEM is given based on the Galerkin formulation (see Praagman, 1979). The principles of the FEM for computational fluid dynamics do not need to be explained in great detail as there are several books available (e.g. Pinder & Gray, 1977, Chung, 1978, Baker, 1983, Fletcher, 1984), most of which, however, do not discuss the SWE.

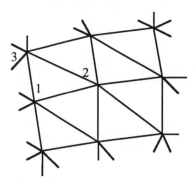

Fig. 6.7. Finite-element mesh; triangular elements

In the FEM, the region is divided into a number of basic elements of (usually) triangular but otherwise irregular shape (fig. 6.7). The unknowns h, u, v are approximated by piecewise smooth functions (usually linear) on each element, i.e. in any point (x,y) you can interpolate linearly between the values at the nodes (the corners of the triangle). In this way the unknowns are defined in the entire region by a sum of piecewise continuous (in this case linear) functions:

$$h(x,y,t) = \sum_{j=1}^{N} h_i(t)\, \phi_i(x,y) \qquad (6.16)$$

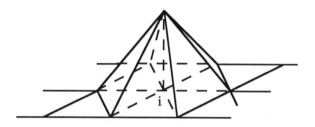

Fig. 6.8. Example of piecewise linear interpolation function

The fields u, v are similarly interpolated, using the same interpolation functions but, of course, different coefficients. The functions ϕ are precisely determined by the following requirements:
- ϕ_i is defined in the entire region
- ϕ_i is linear in each element
- $\phi_i = 1$ in node i, and 0 in all other nodes.

One such function is illustrated in fig. 6.8. As the sum of a number of piecewise linear functions will again be piecewise linear, you get a linear interpolation between nodal values $h_i(t)$. The other unknowns are defined in a similar way. In this form, they will not satisfy the equations; you will be left with residuals which should be forced to get small in some way. This is done by the method of weighted residuals.

Next, the equations to be solved (2.48), (2.51),(2.52) are multiplied by weighting functions $w_j(x,y)$ and integrated over the area. This results in a weighted average residual, which can be made zero for a number of different weighting functions; then it can be assumed (and proved in some cases) that the residual itself will be small. The particular choice in the Galerkin method is $w_j = \phi_j$. One such weighting function is used for any nodal value which is not prescribed by boundary conditions. Consequently, you obtain exactly the correct number of equations for the unknown nodal values.

Applying this operation to the continuity equation (2.48), a system of ordinary differential equations results which can be written formally in matrix form as

$$M_{ij}\frac{dh_j}{dt} + D_{ij}^x(a_k)\, u_j + D_{ij}^y(a_k)\, v_j + D_{ij}^x(u_k)\, a_j + D_{ij}^y(v_k)\, a_j = 0 \qquad (6.17)$$

The matrices of coefficients are built up from contributions of the various elements, such as:

$$M_{ij} = \sum_e M_{ij}^e = \sum_e \int \int_e \phi_i \phi_j \, dx \, dy$$

For one element (fig. 6.7), the definition is as follows. The matrix M^e, called the mass matrix, depends only on the geometry of the element and works on the vector containing the nodal values for that element:

$$M^e = \frac{\Delta}{24} \begin{pmatrix} 2 & 1 & 1 \\ 1 & 2 & 1 \\ 1 & 1 & 2 \end{pmatrix} \tag{6.18}$$

where

$$\Delta = e_2 d_3 - e_3 d_2$$

$$\begin{array}{ll} e_1 = y_2 - y_3 & d_1 = x_3 - x_2 \\ e_2 = y_3 - y_1 & d_2 = x_1 - x_3 \\ e_3 = y_1 - y_2 & d_3 = x_2 - x_1 \end{array}$$

Nodal values other than those from the corners of triangle e do not occur in M^e due to the properties of the interpolation functions ϕ_i. You see that the equation for a certain *nodal value* contains also contributions in the mass matrix (and therefore in the time derivatives) from the other nodes sharing an element with the node considered. The system of ordinary differential equations is therefore coupled already in the time derivatives. This is unpleasant for explicit time integration methods, because it forces you to solve a system of equations in each time step (chapter 8). Sometimes, the *consistent mass matrix* (6.18) is therefore replaced by what is called the *lumped mass matrix*

$$M^e = \frac{\Delta}{6} \begin{pmatrix} 1 & 0 & 0 \\ 0 & 1 & 0 \\ 0 & 0 & 1 \end{pmatrix} \tag{6.19}$$

where the contributions of the neighbouring nodes have been lumped into the central node. This seems a rather harmless operation but it is not, as shown below.

The element matrices D^x and D^y come from the nonlinear terms and therefore contain nodal values, indicated here by v_k. The contributions from element e are:

$$D^{xe}(v_k) = \frac{|\Delta|}{24\Delta} \begin{pmatrix} 2v_1 + v_2 + v_3 \\ v_1 + 2v_2 + v_3 \\ v_1 + v_2 + 2v_3 \end{pmatrix} (e_1 \; e_2 \; e_3)$$

$$D^{ye}(v_k) = \frac{|\Delta|}{24\Delta} \begin{pmatrix} 2v_1 + v_2 + v_3 \\ v_1 + 2v_2 + v_3 \\ v_1 + v_2 + 2v_3 \end{pmatrix} (d_1 \; d_2 \; d_3)$$

Note that these are matrix multiplications, not inner products, so the result is again a 3*3 matrix working on the nodal values of the element.

For the momentum equations, the Galerkin procedure results in

$$M_{ij} \frac{du_j}{dt} + D_{ij}^x(u_k) u_j + D_{ij}^y(v_k) u_j - N_{ij}(f_k)v_j + gD_{ij}^x h_j + N_{ij}(r_k)u_j = 0$$

$$M_{ij} \frac{dv_j}{dt} + D_{ij}^x(u_k) v_j + D_{ij}^y(v_k) v_j + N_{ij}(f_k)u_j + gD_{ij}^y h_j + N_{ij}(r_k)v_j = 0$$

(6.20)

If the derivative operators D^x, D^y are given without argument, the argument is unity in the definition above. Furthermore, for some argument r_k,

$$N^e(r_k) = \frac{|\Delta|}{120} \begin{pmatrix} (6,2,2) & (2,2,1) & (2,1,2) \\ (2,2,1) & (2,6,2) & (1,2,2) \\ (2,1,2) & (1,2,2) & (2,2,6) \end{pmatrix}$$

with the short-hand notation

$$(6,2,2) = 6r_1 + 2r_2 + 2r_3$$

In general, it is not easy to recognize the structure of these equations. However, (for purposes of analysis, not for actual application), it is interesting to see what happens if you choose the grid to be regular. Then the matrices can be specialized and you get a finite-difference formula which can be compared with the previous ones.

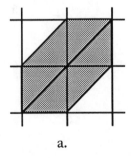
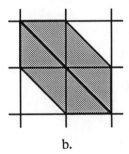

a.   b.

Fig. 6.9. Two ways to triangulate a rectangular grid

On a regular grid, you have a choice how to define the triangular elements. Take the case of fig. 6.9a first. You can assemble all contributions to the central node from the surrounding elements. The operators D^x, D^y are now constants and N coincides with M except in the y-dependent Coriolis term. You get the following finite-difference equations, cast in the same form as those from section 6.2:

$$M_2 \frac{dh}{dt} + UD_{1x}a + aD_{1x}u + aD_{1y}v = 0$$

(6.21)

$$M_2 \frac{du}{dt} + U\, D_{1x}u - (fM_2 + \beta B_y)v + g\, D_{1x}h + rM_2u = 0$$

where the coefficients now indicate the grid stencils (i.e. the coefficients for 3*3 grid points, indicated with square brackets; do not confuse with the matrices in the finite-element equations)

$$M_2 = \frac{1}{24} \begin{bmatrix} 0 & 1 & 1 \\ 1 & 6 & 1 \\ 1 & 1 & 0 \end{bmatrix}$$

$$D_{1x} = \frac{1}{6\Delta x} \begin{bmatrix} 0 & -1 & 1 \\ -2 & 0 & 2 \\ -1 & 1 & 0 \end{bmatrix}$$

$$B_y = \frac{\Delta y}{48} \begin{bmatrix} 0 & 1 & 1 \\ 0 & 0 & 0 \\ -1 & -1 & 0 \end{bmatrix}$$

Clearly, this is a scheme with a directional bias. This can be removed by averaging with the corresponding result from the triangulation shown in fig. 6.9b. Then the grid stencils in eq. (6.21) become (and these are the ones used in the next chapters):

$$M_2 = \frac{1}{24} \begin{bmatrix} 1 & 2 & 1 \\ 2 & 12 & 2 \\ 1 & 2 & 1 \end{bmatrix} \tag{6.22}$$

$$D_{1x} = \frac{1}{12\Delta x} \begin{bmatrix} -1 & 0 & 1 \\ -4 & 0 & 4 \\ -1 & 0 & 1 \end{bmatrix} \tag{6.23}$$

$$B_y = \frac{\Delta y}{48} \begin{bmatrix} 1 & 2 & 1 \\ 0 & 0 & 0 \\ -1 & -2 & -1 \end{bmatrix} \tag{6.24}$$

The other momentum equation gets a similar form. For this regular grid, you can determine the truncation error by developing all terms into Taylor series relative to the central grid point. Then you can check

$$M_2h = h + \frac{1}{6}(\Delta x^2 h_{xx} + \Delta y^2 h_{yy}) + O(\Delta x^4)$$

$$D_{1x}u = u_x + \frac{1}{6}(\Delta x^2 u_{xxx} + \Delta y^2 u_{xyy}) + O(\Delta x^4)$$

If you put this into (6.21) and replace the time derivatives by spatial derivatives using the same equations again, you will notice that all second-order terms exactly cancel and that the method is therefore 4th order accurate in terms of Δx, Δy. The wave-propagation analysis in chapter 7 confirms this. However, for irregular element distributions, it appears that the accuracy drops to 2nd order.

If mass lumping is applied, the grid stencil becomes just

$$M_{2,lumped} = \begin{bmatrix} 0 & 0 & 0 \\ 0 & 1 & 0 \\ 0 & 0 & 0 \end{bmatrix}$$

instead of (6.22). The difference between the consistent and lumped mass matrices is

$$M_{2,cons} - M_{2,\,lumped} = \frac{1}{24} \begin{bmatrix} 1 & 2 & 1 \\ 2 & -12 & 2 \\ 1 & 2 & 1 \end{bmatrix}$$

which, introduced in the x-momentum equation, produces an approximation of

$$\frac{1}{3} \Delta x^2 \nabla^2 \frac{\partial u}{\partial t} \tag{6.25}$$

This is a second-order error, which works only in time-dependent cases. As you will see in the analysis of accuracy (chapter 7), this may have serious consequences for the accuracy with which wave propagation is reproduced. If the same way of lumping is applied in the Coriolis and friction terms of (6.21b) as well, second-order errors arise which persist in the steady state.

6.5. Finite elements for wave equation

The wave equation was proposed by Lynch & Gray (1979) as an alternative to the mass-balance equation in order to get better numerical properties. It seems to have been used only in finite-element form, although this is not at all essential. For the full nonlinear form, see Kinnmark (1986). For purposes of comparison with other methods, it is sufficient to give here the finite-element discretisation of the linear equation (4.6). Using linear interpolation on triangular elements, it is possible to derive

$$M_2 \frac{d^2h}{dt^2} - ga\,(S_{xx} + S_{yy})h + GM_2 \frac{dh}{dt} + (G\text{-}r)\,a\,(D_{1x}u + D_{1y}v) +$$
$$+ af\,(D_{1x}v - D_{1y}u) + a\beta(2B_xv - B_1u) = 0 \tag{6.26}$$

This equation is supposed to be used together with the momentum equations (6.20). Specialized to a regular grid, the grid stencils are (Vreugdenhil, 1990)

$$S_{xx} = \frac{1}{\Delta x^2} \begin{bmatrix} 0 & 0 & 0 \\ 1 & -2 & 1 \\ 0 & 0 & 0 \end{bmatrix}$$

$$B_1 = \frac{\Delta x}{48} \begin{bmatrix} 3 & 2 & 3 \\ 2 & 28 & 2 \\ 3 & 2 & 3 \end{bmatrix}$$

6.6. Grid generation

For finite-element methods as well as for curvilinear finite-difference methods you need some way to generate a grid (or element layout). In the former case, it has already been argued that the grid should be smooth, otherwise your finite-difference expressions will be in error. For finite elements, this appears to be less so on first sight, as the derivation through weighted residuals is set up for an arbitrary layout of the elements. Yet, there are indications that in this case, as well, you need a certain smoothness in the grid. Therefore, the methods discussed in this section may be used in both cases.

The subject of grid generation is a separate field of investigation with its own books (such as the standard work by Thompson et al, 1985) and conferences (e.g. Hauser & Taylor, 1986). Useful review papers are Thompson & Warsi (1982), Eisemann (1985). A short but clear introduction is included in the book by Fletcher (1988). Here, only a few methods are discussed to give an idea of the possibilities. For further information you should consult the references.

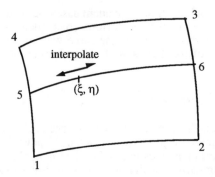

Fig. 6.10. Principle of affine interpolation

The first method, which is relatively simple, yet quite powerful, is affine interpolation (fig. 6.10). A region in the x,y plane is mapped into a square in the ξ,η plane by two successive 1-d interpolations. Any function f (which includes x or y as special cases) can be interpolated along a constant-η line between points 5 and 6 as

$$F_1(\xi, \eta) = (1 - \xi) f_5 + \xi f_6 = (1 - \xi) f(0, \eta) + \xi f(1, \eta) \qquad (6.27)$$

For $\eta = 0$ or 1, this linear interpolation will not coincide with the actual behaviour of $f(\xi,0)$ or $f(\xi,1)$ at the upper and lower boundaries. The next interpolation along lines of constant ξ, distributes the deviations:

$$F(\xi, \eta)) = F_1(\xi, \eta) + (1 - \eta)\ \{f(\xi,0) - F_1(\xi, 0)\} + \eta\ \{f(\xi,1) - F_1(\xi, 1)\} \quad (6.28)$$

F is the final interpolated value. You can check that it produces correct values at each of the boundaries. An example of this technique is given in fig. 6.11. Both x and y values are interpolated between the prescribed boundary values according to (6.28). The technique is limited insofar that you do not have any control on the location of the internal grid points.

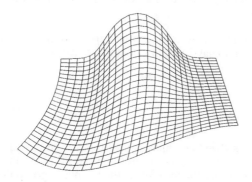

Fig. 6.11. Example of linear affine interpolation

However, there is no obligation to interpolate linearly: the formulae may be generalized by replacing ξ and η on the right-hand side (including the function arguments) of (6.27) and (6.28) by sufficiently smooth and monotonous interpolation functions $\phi(\xi), \psi(\eta)$ with the properties

$$\phi(0) = \psi(0) = 0, \quad \phi(1) = \psi(1) = 1$$

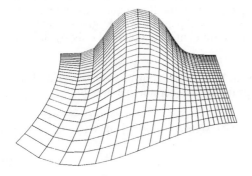

Fig. 6.12. Example of affine interpolation with exponential stretching according to (6.29) with $a = 2.4$

For example, fig. 6.12 has been generated by choosing

$$\phi(\xi) = 1 - \frac{e^{-a\xi} - e^{-a}}{1 - e^{-a}} \qquad (6.29)$$

and the same function for ψ. In this way you can concentrate grid lines near the boundaries or at any other location. Actually, you can control the position of the internal grid points only through stretching of the boundary points. This gives some freedom but it is at the same time a limitation of the technique.

Another way to generate a smooth grid is by letting ξ and η be solutions of elliptic differential equations, which are known to have certain smoothness properties:

$$\begin{aligned} \xi_{xx} + \xi_{yy} &= P(x,y) \\ \eta_{xx} + \eta_{yy} &= Q(x,y) \end{aligned} \qquad (6.30)$$

with "source" or "attraction" functions P and Q which can be used for grid control in the region (see below). Boundary conditions for ξ and η are required all along the closed boundary, which means that you can specify the location of the boundary grid points. The solution of (6.30) will be rather awkward if the region does not have a simple shape in the physical x,y plane, and this is exactly the case in which you would want to do the transformation anyway. Therefore, an essential step in the method is to interchange dependent and independent variables, leading to a system of coupled elliptic differential equations *in the computational ξ,η plane*:

$$g_{22}x_{\xi\xi} - 2g_{12}x_{\xi\eta} + g_{11}x_{\eta\eta} = -J(P\,x_\xi + Q\,x_\eta)$$

$$g_{22}y_{\xi\xi} - 2g_{12}y_{\xi\eta} + g_{11}y_{\eta\eta} = -J(P\,y_\xi + Q\,y_\eta) \qquad (6.31)$$

where the functions g_{ij} are given by (2.58). They depend on x and y with the consequence that (6.31) is a nonlinear system. Even though the equations are more complicated, they can be solved on a simple region, together with the flow equations shown in section 6.6. All computations are done in the computational plane. Fig. 6.13 gives an example with the attraction functions P,Q equal to zero. Grid control is obtained only by the choice of the boundary points.

For many applications, it is useful to have more control over the grid. The attraction functions P,Q can be used to "pull" grid lines to a certain line or point:

$$P(\xi, \eta) = -a\,sign(\xi - \xi_i)\,exp(-c|\xi - \xi_i|) - b\,sign(\xi - \xi_i)\,exp\{-d\,\sqrt{(\xi - \xi_i)^2 + (\eta - \eta_i)^2}\}$$

$$(6.32)$$

or combinations of such terms with suitably chosen parameters $a...d$. Q can be chosen in a similar way with ξ and η interchanged. Figs. 6.13b and 6.14 give examples. The experience is that it is rather cumbersome to find the right attraction functions in order to get the grid look like you want to have it. Moreover, if the source points or lines are within the area, smoothness of the grid lines may be hampered. Johnson & Thompson (1986) applied the method in case of the shallow water equations to get a grid related to the water depth.

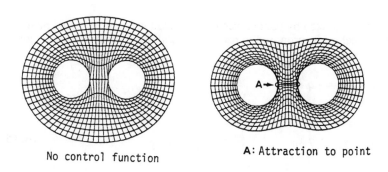

No control function   A: Attraction to point

Fig. 6.13. Grid generated from elliptic equation without and with attraction functions (from Thompson et al, 1985)

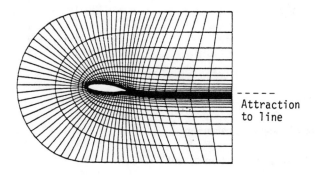

Attraction to line

Fig. 6.14. Example of grid attraction to a line (from Thompson et al, 1985)

Other ways of grid control have been proposed which try to establish a direct link between the grid control functions and the desired grid properties. In fig. 6.15, two such properties are indicated: the aspect ratio and the grid angle, which can be expressed as (Vreugdenhil, 1991)

$$\frac{\Delta n}{\Delta s} = \frac{\Delta \xi}{\Delta \eta} \left(\frac{x_\xi^2 + y_\xi^2}{x_\eta^2 + y_\eta^2} \right)^{\frac{1}{2}}$$

(6.33a)

$$\cos \phi = - \frac{x_\xi x_\eta + y_\xi y_\eta}{\left((x_\xi^2 + y_\xi^2)(x_\eta^2 + y_\eta^2) \right)^{\frac{1}{2}}}$$

(6.33b)

Here, $\Delta \xi$ and $\Delta \eta$ will be fixed by your choice of the grid in the computational plane. Arina (1986), for orthogonal grids, proposed to use the generalized Cauchy-Riemann equations with a control function q:

$$\eta_y = q \, \xi_x, \quad \eta_x = - q \, \xi_y$$

(6.34)

or (using 6.17):

$$x_\xi = q \, y_\eta, \quad y_\xi = - q \, x_\eta$$

(6.35)

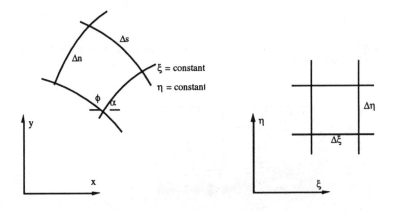

Fig. 6.15. Definition of grid-control parameters

Using this, you will find $\phi = \pi/2$ (as it should) and

$$\frac{\Delta n}{\Delta s} = q \frac{\Delta \xi}{\Delta \eta}$$

The function q therefore determines the aspect ratio of the grid and, through it, the distribution of grid spacing. By making a sketch of the desired grid, you can predict the value of $q(x,y)$ to be used; then x and y can be found by solving the elliptic equation

$$(q^{-1} x_\xi)_\xi + (q \, x_\eta)_\eta = 0$$

(6.36)

which is also satisfied by y but with different boundary conditions. Note that these equations are nonlinear and coupled by the coefficient $q(x,y)$. Boundary conditions are (i) that the grid points should be on the physical boundary, and (ii) that the grid lines be orthogonal at the boundary, i.e. one of the relations (6.35). The exact location of the boundary points cannot be prescribed, but if you have specified the function q correctly, they will be approximately where you want them. If necessary, you can slightly change q and repeat the procedure. Note that the grid will be smooth (second derivatives are continuous) if the function q is continuous (i.e. does not have jumps). An example by Ogink et al (1986), who made the same proposal, is given in fig. 1.6.

Vreugdenhil (1991) extended the procedure to the non-orthogonal case by assuming two control functions $p(x,y)$, $q(x,y)$ and replacing (6.57) by

$$\begin{pmatrix} y_\xi \\ x_\xi \end{pmatrix} = \begin{pmatrix} -p & q \\ -q & -p \end{pmatrix} \begin{pmatrix} y_\eta \\ x_\eta \end{pmatrix} \tag{6.37}$$

Then introducing this into (6.33)

$$\frac{\Delta n}{\Delta s} = \sqrt{p^2 + q^2}\, \frac{\Delta \xi}{\Delta \eta}$$

$$\cos \phi = \frac{p}{\sqrt{p^2 + q^2}} \tag{6.38}$$

and you can control both properties simultaneously by properly choosing the two functions p and q. The grid locations have now to be solved from the elliptic equation

$$\left(\frac{1}{q} x_\xi + \frac{p}{q} x_\eta \right)_\xi + \left(\frac{p}{q} x_\xi + \frac{p^2 + q^2}{q} x_\eta \right)_\eta = 0 \tag{6.39}$$

which is satisfied by y as well. Boundary conditions are similar to the previous case. An example is given in fig. 6.16.

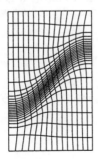

Fig. 6.16. Example of grid refinement using control functions (from Vreugdenhil, 1991)

6.7. Spectral methods

In some fields of application of the SWE or similar equations, particularly in weather forecasting, the spectral method of discretisation has been used extensively, because of its potential to give a very high accuracy with a limited number of degrees of freedom. For an extensive general discussion, see Canuto et al (1988) and for meteorological applications Haltiner & Williams (1980). In the spectral method, the unknown functions are approximated using series of smooth functions. Most commonly Fourier series or Chebyshev polynomials are used; on a sphere, spherical harmonics or Legendre polynomials. There are some serious limitations to the applicability, which prevent use of the method in more general cases:

(i) Fourier series produce periodic functions by definition and therefore work well if the boundary conditions are periodic, i.e. if function values and normal derivatives are equal on both sides of the region; this is the case for (global) atmospheric models. If the boundary conditions are not periodic, it has been tried to simulate them using artificial sources or sinks near the boundaries, but this is a questionable procedure. A better approach is to use Chebyshev polynomials.

(ii) The region should be rectangular, which is not normally the case in SWE applications. However, this problem is not present when boundary-fitted coordinate systems are used (section 6.3), where you have rectangular regions in the computational plane. An extension is the spectral element method, in which the region is split up into relatively coarse quadrilateral elements, within which the spectral method is used.

Various forms of the spectral method can be derived from the method of weighted residuals as in section 6.4. The unknowns are approximated by a series of standard functions like (6.16) but with smooth interpolation functions. The residual R, obtained by introducing the approximation into the differential equations, is weighted with a function w and required to be zero:

$$\int_A R(x,y,t)\, w(x,y)\, dx\, dy = 0$$

which results in ordinary differential equations for the coefficients $h_i(t)$. The choice of interpolating functions is still free. Often, eigenfunctions of a relevant operator are used, e.g. the Laplacian. The reason of this becomes clear below. It is also desirable for the functions to be orthogonal. If you use Fourier series on a region with dimensions L_x, L_y, the interpolating functions are

$$\phi_{m,j}(x,y) = exp\{\, 2\pi i(m\,\frac{x}{L_x} + j\,\frac{y}{L_y})\} \qquad (m = 0...M, \quad j = 0...J) \qquad (6.40)$$

The weighting functions are their complex conjugates (indicated by *). The functions are then orthogonal in the sense that

$$\frac{1}{L_x L_y}\int_0^{L_x}\int_0^{L_y} \phi_{m,j}\phi_{k,n}^* dx\, dy = \delta_{m,k}\delta_{j,n}$$

In weather forecasting on a sphere, the corresponding functions are spherical harmonics (Jarraud & Baede, 1985, or in more detail Jarraud & Simmons, 1983). Another possibility is to use Chebyshev polynomials (Peyret & Taylor, 1983, Fletcher, 1984, 1988). Also, pseudospectral methods may be used where the weight functions are deltafunctions in grid points. This means that the differential equation is enforced in the grid points and the method is a collocation method. Here, only the Fourier Galerkin method is discussed as it seems to be most common for the SWE.

Each unknown function is written as a (generalized) Fourier series

$$h(x,y,t) = \sum_{m=1}^{M} \sum_{j=1}^{J} h_{mj}(t) \; \phi_{mj}(x,y)$$

and similarly for u and v. Note that h_{mj} is not a grid point value as in finite-difference or finite-element methods, but a spectral component. The residuals of the SWE are multiplied by $\phi^*_{k,n}$ and integrated over the (rectangular) region. This gives in principle a system of $3MJ$ equations with the same number of unknowns. Due to the orthogonality property , a nonzero result is obtained only if $k = m$ and $n = j$ (at least for linear systems such as (4.1)):

$$\frac{du_{kn}}{dt} + ik_1U \, u_{kn} - f \, v_{kn} + ik_1g \, h_{kn} + r \, u_{kn} = 0 \qquad (6.41a)$$

$$\frac{dv_{kn}}{dt} + ik_1U \, v_{kn} + f \, u_{kn} + ik_2g \, h_{kn} + r \, v_{kn} = 0 \qquad (6.41b)$$

$$\frac{dh_{kn}}{dt} + ik_1U \, h_{kn} + a\{ \; ik_1 \, u_{kn} + ik_2 \, v_{kn}\} = 0 \qquad (6.41c)$$

where

$$k_1 = 2\pi \, k/L_x, \qquad k_2 = 2\pi \, n/L_y \qquad (6.42)$$

are the wave numbers in x,y directions and $k = 0...M$, $n = 0...J$. The equations for the spectral components turn out to be decoupled in this linear system. If the nonlinear terms are included, this is no longer true; for example the term $u \, \partial u/\partial x$ gives

$$\frac{1}{L_xL_y}\int_0^{L_x} \int_0^{L_y} u \frac{\partial u}{\partial x} \phi^*_{k,n}dx \, dy =$$

$$= \sum_{m,j} \sum_{p,q} \frac{2\pi ip}{L_x} u_{m,j} u_{p,q} \frac{1}{L_xL_y}\int_0^{L_x} \int_0^{L_y} \phi_{m,j}\phi_{p,q} \, \phi^*_{k,n}dx \, dy = \qquad (6.43)$$

$$= \sum_{m,j} \frac{2\pi i \, (k-m)}{L_x} u_{m,j} u_{k-m, \, n-j}$$

The last step is based on the fact that the integral is again nonzero only if $m + p - k = 0$ and $j + q - n = 0$. For each component k,n, you obtain a convolution sum involving MJ operations, so the total amount of work will be of the order $(MJ)^2$ which is very unfavourable. The reason of existence of these convolution sums is just that two functions are multiplied in physical space. A positive property is that care is taken of higher harmonics formed by interaction between components. There is no cut-off in spectral space and no aliasing approximation involved.

Similarly, a variable Coriolis parameter f will lead to a convolution sum:

$$\int_0^\pi \int_0^\pi (f + \beta y)\, exp\{i(mx+jy-kx-ny)\}\, dx\, dy = \delta_{km}\left((f + \tfrac{1}{2}\,\pi)\delta_{jn} + \beta \frac{(-1)^{j-n}}{i\,(j-n)}\right)$$

so the for example the x-momentum equation (6.41a) becomes

$$\frac{du_{kn}}{dt} + ik_1 U\, u_{kn} - (f + \tfrac{1}{2}\pi\beta)v_{kn} + \beta\sum_{j \neq n} \frac{(-1)^{j-n}}{i\,(j-n)}\, v_{kj} + ik_1 g\, h_{kn} + r\, u_{kn} = 0$$

$$(6.44)$$

in which a coupling is found between all the Fourier modes.

You could avoid the convolution procedure, which works fully in spectral space, by performing the multiplications in the nonlinear terms in physical space, because it is a simple procedure there. On the other hand, derivatives are very easy to evaluate in spectral space. In order to perform each operation most conveniently, you will have to (Fourier-) transform back and forth between physical and spectral space. Given the spectral coefficients (such as u_{kn}), spectral coefficients for the derivative $\partial u/\partial x$ are obtained by simply multiplying by ik_1. Next, grid point values for u and $\partial u/\partial x$ are computed by a (2-d) Fast Fourier Transform (FFT) from spectral to physical space. Then the product $u\, \partial u/\partial x$ is computed by simple multiplication in the grid points. Finally, the product is transformed into spectral space by another FFT. Other nonlinear terms are treated the same way. Eqs. (6.41) are used to compute the spectral coefficients at the next time step (chapter 8) and the procedure is repeated. The method hinges on efficient FFT techniques, which are available for Fourier series but less so for spherical harmonics. Even then the transform method has been shown to work more efficiently than the convolution method. However, there is now an aliasing problem, because the Fourier transform of the product $u\, \partial u/\partial x$ will take only the standard MJ terms into account, thereby cutting off the higher harmonics involved in this term. In some cases, for this reason additional terms in the FFT have been taken into account (see further Canuto et al, 1988).

In actual meteorological practice (Haltiner & Williams, 1980 from which the following discussion is taken), the primitive (u, v, h) equations are replaced by equations for vorticity ζ (3.7) and divergence $D = u_x + v_y$. This has some advantages in satisfying the overall vorticity and mass balances. The latter is obtained by taking the divergence of (2.51) and (2.52). If there are no external forces and bottom friction is assumed to be linear (as in (4.1)) for simplicity, you find

$$\frac{\partial D}{\partial t} - \frac{\partial}{\partial y} \{(\zeta + f) u\} + \frac{\partial}{\partial x} \{(\zeta + f) v\} - \{\frac{\partial^2}{\partial x^2} + \frac{\partial^2}{\partial y^2}\}\{gh + \frac{1}{2}(u^2 + v^2)\} + rD = 0$$

(6.45)

You can split the velocity vector into rotational and divergent parts:

$$u = \psi_y + \chi_x \qquad v = -\psi_x + \chi_y$$

(6.46)

Then the relations between the stream function ψ, the potential χ and the variables ζ and D are:

$$\zeta = \nabla^2 \psi \qquad D = \nabla^2 \chi$$

(6.47)

These equations reduce to simple algebraic relations between the spectral components, which explains the idea to use eigenfunctions of the Laplacian as basis functions. For the purpose of comparison, the linear forms are given here (with $U = V = 0$)

$$\frac{\partial h}{\partial t} + a \nabla^2 \chi = 0$$

$$\frac{\partial D}{\partial t} + f \zeta + \beta u - g \nabla^2 h + r D = 0$$

(6.48)

$$\frac{\partial \zeta}{\partial t} + \beta v + f \nabla^2 \chi + r \zeta = 0$$

Expressing everything in terms of h, ψ and χ, you obtain three equations with three unknowns:

$$\frac{\partial h}{\partial t} + a \nabla^2 \chi = 0$$

$$\frac{\partial}{\partial t} \nabla^2 \chi + f \nabla^2 \psi + \beta (\psi_y + \chi_x) - g \nabla^2 h + r \nabla^2 \chi = 0$$

(6.49)

$$\frac{\partial}{\partial t} \nabla^2 \psi + f \nabla^2 \chi + \beta (\chi_y - \psi_x) + r \nabla^2 \psi = 0$$

(6.49c)

For details see Haltiner & Williams. The Fourier-Galerkin procedure can again be applied to this system. If the region of interest is small, the Coriolis parameter f may be assumed constant, as its variation is already accounted for explicitly by the β- terms. Using orthogonality, you find the spectral equations:

$$\frac{dh_{kn}}{dt} - a k^2 \chi_{kn} = 0$$

$$k^2 \left(\frac{d\chi_{kn}}{dt} + f \psi_{kn} - g h_{kn} + r \chi_{kn}\right) + i\beta (k_2 \psi_{kn} + k_1 \chi_{kn} = 0$$

(6.50)

$$k^2 \left(\frac{d\psi_{kn}}{dt} + f\chi_{kn} + r\,\psi_{kn} \right) + i\beta\,(k_2\chi_{kn} - k_1\psi_{kn}) = 0$$

where the coefficient (not the subscript) $k^2 = k_1^2 + k_2^2$ and $k_{1,2}$ are defined in (6.42). As in (6.41), no coupling between different components is found, but again the nonlinear terms will produce a coupling. The velocity components can be found in spectral space from

$$u_{kn} = ik_2\,\psi_{kn} + i\,k_1\,\chi_{kn} \qquad v_{kn} = -\,ik_1\,\psi_{kn} + i\,k_2\,\chi_{kn} \qquad (6.51)$$

In chapter 7 you will see that in its linear form, the spectral method does not introduce any errors due to spatial discretisation. Of course, this is true only for the resolved components. The error is determined therefore by the truncation of the Fourier series. The number of components required to obtain a certain accuracy is determined by the scale of the phenomenon to be studied, that is of the topography and the flow pattern, including possible steep gradients.

Chapter 7

Effects of space discretization on wave propagation

7.1. Harmonic wave propagation

The solution of the discretized equations derived in ch. 6 will not be the same as the continuous one, but hopefully close to it. The question of numerical accuracy can be approached in a number of ways: analysis of the truncation error, experimental numerical convergence analysis and wave-propagation analysis for the linearized equations.

The truncation error has been given in ch. 6 where possible. It gives an indication of what happens to the error if you reduce the grid-size (increase the resolution). Moreover, it shows what sort of terms are actually added to the equation you are trying to solve and these terms can sometimes be interpreted in a quasi-physical sense, e.g. as numerical diffusion or viscosity. This is certainly valuable but incomplete information. The difficulty is that the truncation error is the error in the *equation*, not in the solution, so it is an inverse measure of accuracy. It says to what equation the computed results are a solution. There is a theoretical relation between the truncation error and the error in the solution (the discretization error) which says that in general their orders will be equal, but it difficult to use this relation to estimate the magnitude of the discretization error. The great advantage of truncation-error analysis is that it can be done for nonlinear cases (though usually with a lot of effort).

A basic experimental method to assess the discretization error in the solution is to compute the solution with different step-sizes and observe whether any changes occur. If they do, you will have to do a further refinement. If they don't, you have an indication that the solution is accurate to the number of figures that do not change. Unfortunately, in actual practice it is often out of question to do such an experimental numerical convergence analysis due to the sheer cost. You should realize that halving the grid size means that you get four times the number of grid points. Moreover, you will usually need to halve the time step as well (see ch. 8 and 9), so the amount of work is 8 times that of the basic run and this may well exceed your computer resources, not to speak of a further refinement by a factor of 2.

A partial indication of accuracy in the *solution* (discretization error) can be obtained by considering harmonic wave propagation for the linearized equations. This is discussed in this chapter for the semi-discretized equations and in ch. 9 for the time discretization. As many shallow-water problems do involve waves, this type of analysis can give information of practical significance. The main drawback is that nonlinear effects cannot be accomodated. In many cases these are relatively small so that the linear results are still good indications of the order of magnitude of the numerical errors.

The analysis of accuracy by harmonic wave propagation is not new. It appears to originate from meteorology. See, e.g. Schönstadt (1977) or Jancic & Mesinger (1983). It has also been used by Leendertse (1967) and various other people, however usually only for gravity waves. A very interesting general discussion for hyperbolic equations can be found in Vichnevetsky & Bowles (1982). In case of 3-d flows, a splitting in modes can be made, each of which behaves like the 2-d SWE (section 2.5.2). Song & Tang (1993) investigated numerical errors for the 3-d case in this way.

The approach is the same as in section 4.4: you assume a wavelike solution to the semi-discrete equations:

$$\mathbf{v}_{k,j}(t) = \mathbf{v}_0 exp\{i(k_1 k \Delta x + k_2 j \Delta y - vt)\} \qquad (7.1)$$

The wave length and direction are fixed by the initial condition and therefore the same as in the continuous case. They will not change in with time as the model is linear. However, frequency and amplitude will generally be different from those in the continuous case. These are exactly the two quantities characterizing the wave if you absorb the phase into the (complex) amplitude. Substituting (7.1) into (6.3), for example, gives, after cancelling common factors:

$$\left(- iv + A \frac{e^{i\xi} - e^{-i\xi}}{2\Delta x} + B \frac{e^{i\eta} - e^{-i\eta}}{2\Delta y} + C\right) \mathbf{v}_0 = 0$$

or

$$\left(- iv + ik_1 A \frac{sin\ \xi}{\xi} + ik_2 B \frac{sin\ \eta}{\eta} + C\right) \mathbf{v}_0 = 0 \qquad (7.2)$$

where $\xi = k_1 \Delta x$ and $\eta = k_2 \Delta y$. This is quite similar to (4.15) for the continuous case but there are two differences which constitute the numerical errors:

1. The frequency (eigenvalue) v differs from ω, which means that the phase speed and wave damping of the numerical wave are different from their theoretical values;

2. Generally, the matrices A, B and C do not have common eigenvectors, so the eigenvector $\mathbf{v}_0$ occurring in (7.2) differs from $\mathbf{v}_1$ in (4.15). The eigenvector contains amplitudes and phases of u, v, h with respect to one another, so these relations have changed.

If ξ, η are small, the numerical factors $(sin\ \xi)/\xi$ and $(sin\ \eta)/\eta$ approach unity and you get (4.15) again. This indicates that the numerical accuracy depends on ξ and η to be small. This is so if Δx and Δy are small *relative to the wave length*, as you would expect.

More specifically, you can define the following parameters to indicate the accuracy in physical terms, keeping in mimd that ω and v may be complex:

(i) damping factor = ratio of numerical to exact amplitude after a relevant time t (taken to be the wave period):

$$d = \left| \frac{e^{-ivt}}{e^{-i\omega t}} \right| = \frac{e^{Im(v)t}}{e^{Im(\omega)t}} \tag{7.3}$$

(ii) relative phase or propagation speed

$$c_r = \frac{Re(v)}{Re(\omega)} \tag{7.4}$$

Both should be as close to unity as you need for your particular application so $(1 - d)$ and $(1 - c_r)$ are indications of the error. These are the two quantities we will mainly concentrate on, for the four types of waves identified in ch. 4. It is possible to define similar quantities for ratios of amplitudes and phases of velocity and water-level (the eigenvector components) (Vreugdenhil, 1989), but this not shown here. Eqs. (7.3) and (7.4) are indication of the errors for a specific choice of numerical method and grid sizes. Conversely, if you set a certain level of accuracy, you can use them to select values for the grid sizes that will meet this. Note that it is usually not obvious at all what accuracy you would *require* for a certain application.

Table 7.1. Fourier transforms of finite-difference operators

Definition: $\xi = k_1 \Delta x$, $\eta = k_2 \Delta y$.
Operators in y direction are obtained by straightworward exchange of parameters.

$$D_{0x} \rightarrow \frac{i}{\Delta x} \sin \xi$$

$$D_{1x} \rightarrow \frac{i}{3\Delta x} \sin \xi (2 + \cos \eta)$$

$$M_x \rightarrow \cos \xi$$

$$M_1 \rightarrow \frac{1}{2}(\cos \xi + \cos \eta)$$

$$M_2 \rightarrow \frac{1}{3} + \frac{1}{6}(1 + \cos \xi)(1 + \cos \eta)$$

$$S_{xx} \rightarrow \frac{2}{\Delta x^2}(\cos \xi - 1)$$

$$B_x \rightarrow \frac{i\Delta x}{12} \sin \xi (\cos \eta + 1)$$

$$B_1 \rightarrow \frac{\Delta x}{36} \{(3 \cos \xi + 1)(3 \cos \eta + 1) + 20\}$$

One special case can immediately be recognized from (7.2): if you take $\xi = \eta = \pi$ and omit the zero-order terms Cv for a moment, you find $v = 0$, which means that the waves do not propagate at all; neither are they subject to any damping. Such waves are represented by two points per wave length in each direction separately, which is the theoretical minimum. Such poorly resolved waves apparently are not reproduced with any accuracy. If $Cv \neq 0$, you get either stationary, damped waves (with friction $r \neq 0$, but no Coriolis acceleration $f = 0$), or undamped inertial oscillations ($r = 0, f \neq 0$), neither of which represents the correct physics. The methods involving staggered grids do not support these waves as shown in fig. 6.3, so the problem of stationary "$2\Delta x$ waves" does not occur there. Actually, this is one of the arguments to use staggered grids. Yet, the smallest resolvable wave lengths on such grids, as well, will have a poor numerical accuracy.

As shown in (7.2), the finite-difference operators are replaced by their Fourier transforms. For the various operators defined in ch. 6, the transforms are given in table 7.1 (some operators are also included which are defined later). If two operators are applied consecutively, their Fourier transforms are multiplied, as shown by the following example.

$$D_{0x}h = \frac{e^{i\xi} - e^{-i\xi}}{2\Delta x} h_0 exp\{i(k_1 k\Delta x + k_2 j\Delta y - vt)\} = \frac{i \sin \xi}{\Delta x} h_0 exp\{i(k_1 k\Delta x + k_2 j\Delta y - vt)\}$$

$$M_y D_{0x}h = \frac{e^{i(\xi+\eta)} - e^{-i(\xi-\eta)} + e^{i(-\xi+\eta)} - e^{-i(\xi+\eta)}}{4\Delta x} h_0 exp\{i(k_1 k\Delta x + k_2 j\Delta y - vt)\} =$$

$$= \cos \eta \, \frac{i \sin \xi}{\Delta x} h_0 exp\{i(k_1 k\Delta x + k_2 j\Delta y - vt)\}$$

7.2. Gravity waves

Specialising (7.2) to the case of gravity waves with Coriolis effect in a moving reference frame (4.22) gives a set of homogeneous equations, the determinant of which should be zero:

$$\begin{vmatrix} -iv & -f & ik_1 g \dfrac{\sin \xi}{\xi} \\[3mm] f & -iv & ik_2 g \dfrac{\sin \eta}{\eta} \\[3mm] ik_1 a \dfrac{\sin \xi}{\xi} & ik_2 a \dfrac{\sin \eta}{\eta} & -iv \end{vmatrix} = 0 \tag{7.5}$$

which gives the eigenvalues $v_1 = 0$ (which represents the vorticity wave and is discussed in section 7.3) and two values corresponding to the gravity waves:

$$v_{2,3} = \pm \sqrt{ga} \left(\frac{sin^2 \, \xi}{\Delta x^2} + \frac{sin^2 \, \eta}{\Delta y^2} + \frac{f^2}{ga} \right)^{\frac{1}{2}} \tag{7.6}$$

This corresponds to an undamped wave, so the damping factor d is exactly 1 (no amplitude error). The frequency can be compared to its continuous counterpart in (4.21). The relative propagation speed is found to be:

$$c_r = \left(\frac{sin^2 \, \xi}{k^2 \Delta x^2} + \frac{sin^2 \, \eta}{k^2 \Delta y^2} + (Rk)^{-2} \right)^{\frac{1}{2}} \{ 1 + (Rk)^{-2} \}^{-\frac{1}{2}} \tag{7.7}$$

where R is again the Rossby radius. This expression is valid for the unstaggered A-grid. Fig. 7.1. shows the behaviour of this ratio as a function of ξ for the special case where $k_1 = k_2$ (other cases do not produce essentially different results). Note the logarithmic scales. It is seen that the error decreases quadratically with ξ which means that the error in the *solution* is $O(\Delta x^2)$. This agrees with the second-order truncation error found in (6.4).

Introducing the advective terms just means that v in (7.5) is replaced by

$$v - k_1 U \frac{sin \, \xi}{\xi} - k_2 V \frac{sin \, \eta}{\eta} \tag{7.8}$$

which represents the numerical error in advection. This occurs in all three roots and is discussed in section 7.3

On the Arakawa-E grid, (6.6) shows that you get the same result as for the unstaggered A-grid except for the advective terms. Therefore, in fig. 7.1 which does not include advection, the curves for these two grids coincide. On the C-grid, however, you do get a difference due to the averaging in the Coriolis terms (6.8). Then, the eigenvalue problem becomes

$$\begin{vmatrix} -iv & -f \cos \xi \cos \eta & ik_1 g \dfrac{\sin \xi}{\xi} \\ f \cos \xi \cos \eta & -iv & ik_2 g \dfrac{\sin \eta}{\eta} \\ ik_1 a \dfrac{\sin \xi}{\xi} & ik_2 a \dfrac{\sin \eta}{\eta} & -iv \end{vmatrix} = 0$$

so the gravity-wave solution is found as

$$c_r = \left(\frac{sin^2 \, \xi}{k^2 \Delta x^2} + \frac{sin^2 \, \eta}{k^2 \Delta y^2} + \left(\frac{cos \, \xi \, cos \, \eta}{Rk} \right)^2 \right)^{\frac{1}{2}} \{ 1 + (Rk)^{-2} \}^{-\frac{1}{2}} \qquad (7.9)$$

which gives a decrease of accuracy in fig. 7.1: the errors are larger than for the A and E grids to such an extent that you would have to refine the C grid by a factor of 2 to get the same accuracy. Of course, this depends strongly on the importance of the Coriolis term (see section 4.4.1): for $f = 0$, the three grids give exactly the same accuracy.

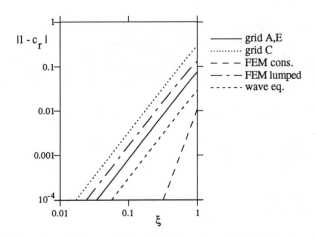

Fig. 7.1.Relative error in wave speed for gravity waves (for $Rk = 1$)

Finite elements can be studied only in a regular form as given by (6.21)...(6.24). For the consistent-mass case, you find in the same way as above

$$v = \pm \sqrt{ga} \left(- \frac{D_{1x}^2 + D_{1y}^2}{M_2^2} + \frac{f^2}{ga} \right)^{\frac{1}{2}} \qquad (7.10)$$

where the operators should be taken from table 7.1. The lumped-mass case is similar but with $M_2 = 1$ substituted. Both cases are shown in fig. 7.1. The consistent mass FEM is more accurate than the grid methods; actually, it shows a 4th order convergence. However, it is unlikely that this applies to irregular finite-element meshes as well (section 7.7). The lumped-mass case, anyway, loses this 4th order convergence and is roughly comparable in accuracy to grids A and E at the same grid size.

Finally, for the wave-equation FEM formulation (section 6.5), the procedure gives a rather complicated result. The eigenvalues involve the expression

$$D_{1x}^2 + D_{1y}^2 - M_2(S_{xx} + S_{yy})$$

which would be zero in continuous form but does not vanish in discrete form. The difference is to lowest order $(k\xi)^2/4$, which has been used in the following expression:

$$v^2 = \frac{1}{2M_2}\left(-\{ga(S_{xx}+S_{yy}) - f^2M_2\} \pm \{(ga(S_{xx}+S_{yy}) - f^2M_2)^2 - ga\,f^2k^2\xi^2\}^{1/2}\right) \quad (7.11)$$

where $G = r = 0$ is assumed (see discussion in section 4.4.5). There are now four roots, one of which is spurious due to the higher order of (6.26). The - sign gives the gravity waves. Fig. 7.1. shows that the wave equation FEM is relatively accurate for gravity waves, though it does not reach the 4th order convergence of conventional FEM. Actually, a variant of the method might be interesting in which the second derivatives cancel exactly, by deriving the wave equation in semi-discrete rather than in continuous form.

The numerical propagation (phase) speed of harmonic waves generally depends on the wave number, even if their continuous counterpart does not: the waves are dispersive and it makes sense to look at the group velocity. For Poincaré waves, discussed in this chapter, the numerical group velocity on A and E grids can be defined from (7.6) as

$$\mathbf{c}_g = \left(\frac{\partial v}{\partial k_1}, \frac{\partial v}{\partial k_2}\right) = \pm\sqrt{ga}\left(\frac{sin^2\,\xi}{\Delta x^2} + \frac{sin^2\,\eta}{\Delta y^2} + \frac{f^2}{ga}\right)^{-\frac{1}{2}}\left(\frac{sin\,\xi\,cos\,\xi}{\Delta x}, \frac{sin\,\eta\,cos\,\eta}{\Delta y}\right) \quad (7.12)$$

The ratio to the group velocity in the continuous case (section 4.4.1) is (componentwise)

$$\frac{c_{g1\,num}}{c_{g1\,ex}} = \frac{1}{c_r}\frac{sin\,2\xi}{2\xi}, \quad \frac{c_{g2\,num}}{c_{g2\,ex}} = \frac{1}{c_r}\frac{sin\,2\eta}{2\eta} \quad (7.13)$$

with c_r according to (7.7). The error in the group velocity is composed of two parts. One comes from the factor $1/c_r$ which indicates the error in the phase velocity but in the opposite way: if the phase speed is too small, the group velocity is too large and conversely. The other part comes from the factor $(sin\,2\xi)/2\xi$. For small ξ, this deviates from unity by $O(\Delta x^2)$, but at larger ξ, important deviations occur and you even get a change of sign at $\xi = \pi/2$. For such short waves (less than 4 points per wavelength), the energy will therefore be propagating in the wrong direction. The possible effects of this have been amply demonstrated by Vichnevetsky (1987) and Trefethen (1982) with the conclusion that you should avoid having signficant waves in this range. This again stresses that ill-resolved waves cannot be taken seriously in a numerical simulation. In the other discretizations, you find a similar effect, though in more complicated form.

7.3. Vorticity waves

For the simplest case of vorticity transport, take $f = r = 0$ so you obtain pure advection of vorticity with the mean flow (4.3). The frequency of travelling waves in continuous form is $\omega = 0$ in the moving coordinate system. The exact frequency in stationary coordinates is

$$\omega = k_1 U + k_2 V$$

In most of the semi-discrete cases, you find a corresponding root $v = 0$ in moving coordinates. However, the behaviour in stationary coordinates may contain errors, due to the discretization of the advection terms. To evaluate these, the equation for advection of vorticity in semi-discrete form is constructed by taking the discrete curl of the momentum equations, such that the pressure gradients are eliminated. For example, on the unstaggered grid A, subtracting D_{0y} times the x-momentum equation from D_{0x} times the y-momentum equation (6.3 with $f = \tilde{r} = 0$) you find

$$\frac{\partial \zeta}{\partial t} + U D_{0x}\zeta + V D_{0y}\zeta = 0 \qquad (7.14)$$

where

$$\zeta = D_{0x}v - D_{0y}u$$

is the discrete vorticity. Inserting a travelling wave solution in (7.14), the frequency for the semi-discrete case is found to be

$$v = k_1 U \frac{\sin \xi}{\xi} + k_2 V \frac{\sin \eta}{\eta} \qquad (7.15)$$

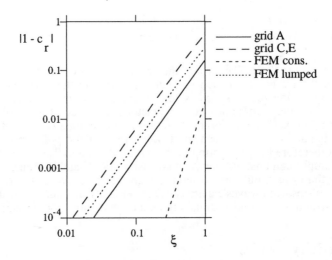

Fig. 7.2. Relative wave speed for advection of vorticity

This is a real value, so there is no numerical wave damping ($d = 1$). The ratio between v and the exact value ω is given in fig. 7.2. for $V=0$. You observe again a quadratic behaviour of the error in terms of ξ, that is, Δx (for a fixed wave number). On grid E, a similar procedure applied to (6.6) results in

$$v = k_1 U \cos \eta \, \frac{\sin \xi}{\xi} + k_2 \, V \cos \xi \, \frac{\sin \eta}{\eta} \qquad (7.16)$$

and grid C gives

$$v = k_1 U \frac{\sin 2\xi}{2\xi} + k_2 \, V \frac{\sin 2\eta}{2\eta} \qquad (7.17)$$

The latter two results happen to be identical if $\xi = \eta$ which has been assumed in fig. 7.2. You can see that grids E and C are less accurate than grid A and to get the same accuracy you need to halve the grid size.

For the consistent FEM, (6.21) result in

$$v = -i\left(U \frac{D_{1x}}{M_2} + V \frac{D_{1y}}{M_2}\right) = k_1 U \frac{\sin \xi}{\xi} \, \frac{2 + \cos \eta}{1 + \frac{1}{2}(1 + \cos \xi)(1 + \cos \eta)} + \text{similar for } y$$

$$(7.18)$$

The lumped-mass FEM is obtained by taking $M_2 = 1$:

$$v = k_1 U \frac{\sin \xi}{3\xi} (2 + \cos \eta) + \text{similar for } y \qquad (7.19)$$

For the wave equation, there is no root $v = 0$ in the moving reference frame. From (7.11), the root with the + sign can be approximated as

$$v^2 \approx \frac{g a f^2 k^2 \xi^2}{4 \, M_2 \{ g a \, (S_{xx} + S_{yy}) - f^2 M_2 \}}$$

which gives the speed of the vorticity wave in a moving frame, i.e. relative to the advection velocity. Making this dimensionless with the gravity-wave speed, you find the relative wave speed

$$\frac{v^2}{g a \, k^2} \approx \frac{\xi^2}{4 \, M_2 \{ \frac{g a}{f^2} (S_{xx} + S_{yy}) - M_2 \}} \approx \frac{\xi^2}{4(1 + R^2 k^2)} \qquad (7.20)$$

which is illustrated in fig. 7.3 for $Rk = 1$. Both the vorticity wave and the spurious wave move with this speed. Note that this comes on top of the discrete advection velocity, which is identical with the consistent FEM, the momentum equations being the same. It is seen that a significant error can be obtained in the behaviour of the vorticity wave.

In the numerical advection speed for all grids, dispersion is introduced: the speed depends on the wave number. A group-velocity effect similar to that discussed in section 7.2 occurs. Again you find a reversal of the group-velocity direction at $\xi = \pi/2$.

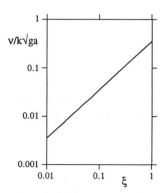

Fig. 7.3. Error in advection velocity for wave-equation method

7.4. Flood waves

In section 4.4.3 it was found that flood waves move with the mean flow (at least in the linearized case), while being damped according to (4.25):

$$\omega = - i \frac{ga\,k^2}{r}$$

The numerical errors can be split into wave propagation errors, which are the same as in the preceding section, and wave damping errors. You have to choose a time at which the amplitudes of numerical and "exact" waves can be compared. A logical choice is the relaxation time at which the wave amplitude is reduced by a factor e^{-1}:

$$exp(Im(\omega)t) = e^{-1}$$

This gives a damping ratio

$$d = \frac{exp(Im(v)t)}{exp(Im(\omega)t)} = exp\{1 - \frac{Im(v)}{Im(\omega)}\} \tag{7.21}$$

which is given in fig. 7.4 for the various discretisations.

For grid A, eqs. (4.26) become

$$ru + g\,D_{0x}h = 0$$
$$rv + g\,D_{0y}h = 0$$
$$\frac{\partial h}{\partial t} + a\{D_{0x}u + D_{0y}v\} = 0 \tag{7.22}$$

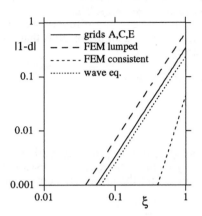

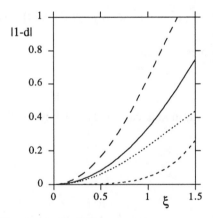

Fig. 7.4. Wave-damping errors for flood waves
top: logarithmic scale, bottom: linear scale

Substituting a harmonic wave solution, the result is

$$v = -i\frac{ga}{r}\left\{\left(k_1\frac{sin\ \xi}{\xi}\right)^2 + \left(k_2\frac{sin\ \eta}{\eta}\right)^2\right\}$$

so

$$d = exp\left[1 - \frac{1}{k^2}\left\{\left(k_1\frac{sin\ \xi}{\xi}\right)^2 + \left(k_2\frac{sin\ \eta}{\eta}\right)^2\right\}\right]$$

(7.23)

On grids E and C, you find exactly the same result, which is not so strange since the terms in (7.22) are treated the same way in all three cases. The convergence is again of second order in Δx. For large Δx, extremely large errors are found as seen in fig. 7.4b which contains the same data but on a linear scale. Again the result is that ill-resolved waves are represented with large numerical errors.

For the consistent FEM on a regular grid, you get

$$v = i \frac{ga}{r M_2^2} \left\{ D_{1x}^2 + D_{1y}^2 \right\} \tag{7.24}$$

which is again 4-th order accurate (but this is probably no longer true if the grid is irregular). For the lumped FEM, you take the same expression with $M_2 = 1$. The convergence is then only second order and the errors are somewhat larger than for the finite-difference methods.

For the wave-equation approach, the momentum equations are the same as for the consistent FEM. With the wave equation (4.26), the frequency follows from

$$(G-r)gar \, (D_{1x}^2 + D_{1y}^2) + r^2 M_2 \, \{ v^2 + ivG \, M_2 + ga(S_{xx} + S_{yy}) \} = 0 \tag{7.25}$$

In the limit for $\xi, \eta \to 0$, this becomes

$$v^2 + ivG - ga \, k^2 \frac{G}{r} = 0$$

which looks very much like the continuous case (4.17). If k is small, the approximate roots are

$$v_1 = -i \, G$$
$$v_2 = -i \frac{ga \, k^2 \, G}{r^2}$$

the former indicating the spurious wave and the latter the flood wave. You see by comparison with (4.25) that you will get the correct damping rate only if $G = r$. This is no longer a more or less arbitrary choice but an obligatory one for the correct reproduction of flood waves. With this choice, the approximate equation for v for finite grid sizes becomes

$$v_2 = i \frac{ga \, (S_{xx} + S_{yy})}{r M_2} \tag{7.26}$$

which is included in fig. 7.4. The wave equation method is found to be about as accurate as finite differences on grids A,C,E.

7.5. Rossby waves

In the continuous case, given by (4.28), the behaviour of Rossby waves is obtained by using the (potential) vorticity equation. This is obtained in two steps: first the pressure is eliminated from the momentum equations (4.22a,b) by cross-differentiation, and secondly, the divergence is eliminated using (4.22c). In the discrete case, similar operations can be performed to get a discrete potential-vorticity equation. The result may be different on various grids. For example, on grids A and E, D_{0x} times the y-momentum equation, less D_{0y} times the x-momentum equation, less (f/a) times the continuity equation gives:

$$\frac{\partial}{\partial t}(D_{0x}v - D_{0y}u - \frac{f}{a}h) + f D_{0x}u + D_{0y}(fv) - f(D_{0x}u + D_{0y}v) = 0$$

Here, the Coriolis parameter f is still variable. You may check by substitution that for any two grid functions a, b

$$D_{0y}(ab) = M_y a\, D_{0y}b + M_y b\, D_{0y}a$$

so with $f = f_0 + \beta y$,

$$D_{0y}(fv) = f_0 D_{0y}v + \beta M_y v$$

and you obtain the semi-discrete potential-vorticity equation (suppressing the subscript on f)

$$\frac{\partial}{\partial t}(D_{0x}v - D_{0y}u - \frac{f}{a}h) + \beta M_y v = 0 \tag{7.27}$$

This should be combined with the momentum equations

$$- fv + g D_{0x}h = 0$$
$$fu + g D_{0y}h = 0$$

where now f may be taken constant, its variation being explicitly accounted for in (7.27). Inserting a harmonic wave solution gives

$$v = i\beta \frac{D_{0x}M_y}{- D_{0x}^2 - D_{0y}^2 + \frac{f^2}{ga}} \tag{7.28}$$

in which the Fourier transforms from table 7.1 should be used. As v turns out to be real, there is no wave damping. The error in the wave speed, which is the ratio to the exact expression (4.29), is illustrated in fig. 7.5. A second-order accuracy is found.

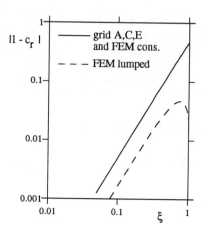

Fig. 7.5. Error in wave speed for Rossby waves

In a similar way, the equations on grid C turn out to be

$$- f M_x M_y v + g D_{0x} h = 0$$
$$f M_x M_y u + g D_{0y} h = 0$$

$$\frac{\partial}{\partial t} (D_{0x} v - D_{0y} u - \frac{f}{a} M_x M_y h) + \beta M_x M_y^2 v = 0$$

(7.29)

from which you find

$$v = i\beta \frac{D_{0x} M_x M_y^2}{- D_{0x}^2 - D_{0y}^2 + f^2 M_x M_y / g a}$$

(7.30)

The result is hardly distinguishable from that for grids A and E. In the finite-element method, a similar procedure gives

$$\frac{d}{dt} (D_{1x} M_2 v - D_{1y} M_2 u - \frac{f}{a} M_2^2 h) + D_{1x} f M_2 u + D_{1y} f M_2 v - f M_2 (D_{1x} u + D_{1y} v) = 0$$

Introducing $f = f_0 + \beta y$ again (with $y = 0$ at grid point k, j), the terms with f_0 outside the time derivative turn out to cancel. You may check by substitution

$$\beta\, D_{1y}yM_2v = \frac{\beta}{6}\, M_y(4 + 2M_x)M_2v$$

$$\beta\, D_{1x}yM_2u = \frac{\beta\, \Delta x^2}{3}\, D_{0x}D_{0y}M_2u$$

so the potential-vorticity equation becomes

$$\frac{d}{dt}\left(D_{1x}M_2v - D_{1y}M_2u - \frac{f}{a}M_2^2h\right) + \frac{\beta}{6}\, M_y(4 + 2M_x)M_2v + \frac{\beta\, \Delta x^2}{3}\, D_{0x}D_{0y}M_2u = 0 \tag{7.31}$$

It is seen that a term involving βu occurs which is not there in continuous form. The momentum equations become

$$-f M_2\, v + g D_{1x}h = 0$$
$$f M_2\, u + g D_{1y}h = 0$$

The wave-propagation analysis now gives

$$v = i\, \beta\, \frac{\frac{1}{6}D_{1x}M_y(4 + 2M_x) - \frac{1}{3}\Delta x^2\, D_{1y}D_{0x}D_{0y}}{-D_{1x}^2 - D_{1y}^2 + M_2^2\, \frac{f^2}{ga}} \tag{7.32}$$

In the lumped-mass FEM, you take $M_2 = 1$ (supposing that you do this also in the Coriolis terms). Fig. 7.5. shows that the FEM is about as accurate as the finite-difference methods in this case. The accuracy is of second order. Mass lumping turns out to have little influence.

For the wave-equation approach, to get the potential-vorticity equation, you should unwrap the wave equation (6.26) into a continuity equation. In the semi-discrete case, it is not at all evident how to do that. The accuracy for Rossby waves can therefore not be studied with this method.

As mentioned in section 6.7, the spectral method gives coupled equations for all modes if a variable Coriolis parameter is introduced. This makes a wave-propagation analysis very difficult. Anyway, the exact reproduction found for the other wave types does not apply here. For an evaluation, numerical experiments will be needed.

Again, for Rossby waves numerical effects in the group velocity can be observed. For example, on the A and E grids (7.28) results in a numerical group velocity

$$c_{g1} = \frac{-\beta \cos \xi \cos \eta}{\left(\dfrac{\sin^2\xi}{\Delta x^2} + \dfrac{\sin^2\eta}{\Delta y^2} + \dfrac{f^2}{ga}\right)^2}\left(\frac{\sin^2\eta}{\Delta y^2} - \frac{\sin^2\xi}{\Delta x^2} + \frac{f^2}{ga}\right) \tag{7.33}$$

$$c_{g2} = \frac{\beta \sin \xi \sin \eta}{\left(\frac{\sin^2\xi}{\Delta x^2} + \frac{\sin^2\eta}{\Delta y^2} + \frac{f^2}{ga}\right)^2} \left(\frac{\sin^2\eta}{\Delta y^2} + \frac{\sin^2\xi}{\Delta x^2} + \frac{2\cos^2\eta}{\Delta y^2}\right)$$

The ratio of this result to the analytical group velocity (4.29) is illustrated in fig. 7.6. Large errors are found for poorly resolved waves.

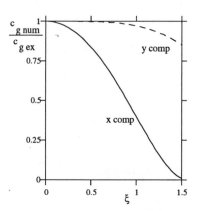

Fig. 7.6. Error in Rossby-wave group velocity for A,E grids

7.6. Rotated grids

On curvilinear grids, the finite-difference equations, even in linearized form, have variable coefficients containing the grid transformation functions. The same types of differencing (staggered or unstaggered grids) can be applied in the computational region. Generally speaking, you may expect to have the same order of accuracy as on Cartesian grids, *provided the grid is sufficiently smooth.*

The wave propagation analysis as discussed in this chapter can not be done for arbitrary grids. There is, however, one question that can at least partially be answered using linear wave-propagation analysis: does it make a difference whether you use Cartesian or contravariant velocity components? In the simplest case, the transformed grid is just rectangular but rotated over an angle ϕ:

$$x = \xi \cos \phi - \eta \sin \phi$$
$$y = \xi \sin\phi + \eta \cos \phi$$

Using contravariant velocity components then yields finite difference equations in the rotated grid identical to the original ones (at least for the linear part). Therefore the same accuracy will be reached. On the other hand, using the Cartesian components on a rotated grid gives the equations (compare (2.55))

$$\frac{\partial \mathbf{v}}{\partial t} + A' \frac{\partial \mathbf{v}}{\partial \xi} + B' \frac{\partial \mathbf{v}}{\partial \eta} + C \mathbf{v} = 0 \tag{7.34}$$

where

$$A' = A \cos \phi + B \sin \phi, \quad B' = - A \sin\phi + B \cos \phi$$

Now suppose that you apply finite differences on an A-type grid in the computational region. The difference equations then read

$$\frac{\partial \mathbf{v}}{\partial t} + A' D_{0\zeta}\mathbf{v} + B' D_{0\eta}\mathbf{v} + C \mathbf{v} = 0$$

Inserting a travelling harmonic wave gives

$$\{- i \, v + A' D_{0\xi} + B'D_{0\eta} + C\} \, v_0 = 0$$

or

$$\{- i \, v + (A \cos \phi + B \sin \phi) \, D_{0\xi} + (- A \sin\phi + B \cos \phi)D_{0\eta} + C\} \, v_0 = 0$$

By comparison with (7.5) it is easy to see that the eigenvalues for the gravity-wave case are

$$v = \pm\sqrt{ga}\left(- (\cos \phi \, D_{0\xi} - \sin \phi \, D_{0\eta})^2 - (\sin \phi \, D_{0\xi} + \cos \phi \, D_{0\eta})^2 + \frac{f^2}{ga}\right)^{1/2} \tag{7.35}$$

But this is identical to

$$v = \pm\sqrt{ga}\left(- D_{0\xi}^2 - D_{0\eta}^2 + \frac{f^2}{ga}\right)^{1/2}$$

which you obtain in the case of contravariant components. You can conclude that there is no difference in wave propagation between the two cases.

7.7. Irregular grids

Irregular grids, in which the grid size does not vary smoothly from one grid interval to the next, are allowed by the finite-element method. The weighted residual formalism ensures that you will get a consistent approximation of the differential equations for any element shape or distribution. However, there may be an influence on the accuracy. The accuracy analysis for wave propagation used in this chapter is only applicable to constant-coefficient equations, which means (among other things) that the grid should be uniform, so it is not possible to analyse the effect of irregular grids in this way. An alternative would be to perform systematic numerical experiments for a case in which you know the exact solution, but this does not seem to have been done.

Kinnmark (1986) gives an interesting example of a non-uniform one-dimensional grid consisting of two alternating grid sizes Δx_1 and $\Delta x_2 = \alpha\,\Delta x_1$. In this special case of an "irregular" grid, a wave-propagation analysis can be performed if you allow for alternating amplitudes between consecutive grid points. In either even or odd grid points, a long wave will result, the accuracy of which can be judged; in the other points a slightly different amplitude will be found, so a small ripple is superimposed on the wave.

The equations resulting from the 1-d FEM are specialized to the gravity-wave case as an example. Details are not given here. An undamped travelling wave is found, the speed of which depends on α and $(1+\alpha)k\Delta x$. The latter can be interpreted as 2ξ in comparison with a uniform grid with the same number of grid points; ξ has therefore been plotted on the horizontal axis of fig. 7.7 so that the curves for various values of the grid-size ratio α correspond to equal numbers of grid points per wave length.

For $\alpha = 1$ (regular grid), a 4th order convergence can be observed which agrees with the results in fig. 7.1 for the 2-d FEM case (consistent mass). However, for other values of α, a second-order error shows up at small values of ξ. Admittedly, the additional error is not very large (even for a high value such as $\alpha = 5$) but the 4th order convergence is lost. It is to be expected that this is also true for more general irregular grids.

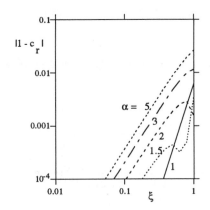

Fig. 7.7. Error in wave speed for gravity wave on irregular grid

7.8. Discrete conservation

In section 3.3, some conservation laws are discussed which follow from the differential equations:
- linear conservation (mass and momentum)
- quadratic conservation (energy and potential enstrophy)

It is not evident that similar properties are valid for the (semi-) discretized equations. You could wonder whether there is any use in trying to satisfy discrete conservation laws.

Linear conservation has been shown to be essential if discontinuities develop in the solutions. This does not happen so commonly in shallow water, but examples are dambreak waves and tidal bores (example in section 1.9). They are similar to shock waves in aerodynamics and have the mathematical character of weak solutions. The Rankine-Hugoniot jump relations are discussed in section 3.4. Lax & Wendroff (1960) have shown that finite-difference methods will "capture" the shocks, that is, produce weak solutions with the correct jump relations (though with a certain amount of "numerical diffusion") under a number of conditions, one of which is that the difference equations satisfy linear conservation of mass and momentum. This is obtained if the equations can be written in the form

$$\frac{\partial a_{i,j}}{\partial t} + \frac{f_{i+\frac{1}{2},j} - f_{i-\frac{1}{2},j}}{\Delta x} + \frac{g_{i,j+\frac{1}{2}} - g_{i,j-\frac{1}{2}}}{\Delta y} = 0 \tag{7.36}$$

where a is any variable and (f,g) its flux. Adding these equations over all grid points results in cancelling of fluxes: the flux leaving one cell equals the flux entering the next cell. The semi-discretization (6.5) satisfies this property if you define

$$f_{i+\frac{1}{2},j} = \frac{1}{2}\{(au)_{i+1,j} + (au)_{i,j}\} \tag{7.37}$$

and similarly for the other terms. Forms like (6.7), (6.9) are based on the non-conservative (advective) form and are therefore not suitable for shock-capturing. However, you can apply the same differencing techniques to (2.49),(2.50) to obtain a linearly conservative method. For a further discussion see, e.g. LeVeque (1990). There is a difference between conservative and non-conservative methods only in the nonlinear case. For the linearized equations (6.6),(6.8) to be used for accuracy analysis, there is no difference.

Except for shock-capturing, the need to satisfy discrete conservation laws is subject to discussion. You could argue that even summation of (7.38) (supposing that it is the mass balance equation) does not yield exact conservation of integrated mass, but rather a numerical (trapezoidal-rule) approximation of it, accurate to second order in Δx, Δy. A nonconservative finite-difference method, the solution of which is a second-order approximation to the true solution, will also produce a total mass, not exactly constant, but accurate to second order. Similar arguments apply to quadratic conservation. Having the total (discrete) energy constant does not mean that you have an accurate solution: the energy may be at wrong places or wrong wave numbers.

On the other hand, some arguments in favor of quadratic conservation have been given as follows. First of all, energy looks somewhat like a quadratic norm for the discrete solution. If this can be shown not to grow unboundedly in time (or even to remain constant), you have a stable solution. This "energy method" for proving stability is used a.o. in the discussion of numerical boundary conditions (section 10.2). It gives more information than the usual linear stability conditions which do not apply to nonlinear equations. There is a difficulty that the energy is not really equivalent to a quadratic norm as there is no guarantee that the depth a will be strictly positive (bounded away from zero), but in many cases it can be safely assumed to be so. Only in the (not unimportant) case of, e.g. flooding and drying of tidal flats and sand banks may this assumption be false.

The importance of enstrophy conservation is that it puts a restriction on the spectral distribution of energy in wave-number space (Arakawa & Lamb, 1981, Sadourny 1975). If $E(k)$ is the spectral energy density at wave number k, then total energy is

$$E_{tot} = \int_0^\infty E(k) \, dk$$

and total enstrophy is

$$Z_{tot} = \int_0^\infty k^2 E(k) \, dk$$

Due to the k^2 factor, small-scale eddies contribute much more to enstrophy than to energy. Now suppose that some energy would be transferred from a certain wave number to a higher one, without affecting total energy. This would cause Z to increase because there is a larger coefficient k^2 in front of the energy density. Therefore, if both E and Z are constant, such an energy transfer should be accompanied by a simultaneous transfer to smaller wave numbers. Effectively, this means that small-scale eddies cannot grow very much if enstrophy is to be conserved. Arakawa & Lamb (1981) argue that this is a desirable feature in numerical methods: it prevents spurious growth of small-scale noise and the energy will remain at larger scales where it is numerically well resolved and reproduced accurately. They give examples of atmospheric flow over a (very crudely resolved) mountain range which illustrate the point.

A classic paper by Arakawa (1966) shows how to obtain full linear and quadratic conservation for a non-divergent 2-d flow for which the vorticity equation is (3.7):

$$\frac{\partial \zeta}{\partial t} + J(\psi, \zeta) = 0 \tag{7.38}$$

$$J(\psi, \zeta) = \psi_x \zeta_y - \psi_y \zeta_x$$

$$u = -\psi_y \quad v = \psi_x$$

where J is the Jacobian and ψ the stream-function (which exists if the flow is non-divergent). Finite-difference approximations of the Jacobian can be constructed in various forms, as explained concisely by Haltiner & Williams (1980):

$$J_1(\psi, \zeta) = \psi_x \zeta_y - \psi_y \zeta_x \approx D_{0x}\psi \, D_{0y}\zeta - D_{0y}\psi \, D_{0x}\zeta$$

$$J_2(\psi, \zeta) = \{\psi\zeta_y\}_x - \{\psi\zeta_x\}_y \approx D_{0x}(\psi \, D_{0y}\zeta) - D_{0y}(\psi \, D_{0x}\zeta) \tag{7.39}$$

$$J_3(\psi, \zeta) = \{\zeta \psi_x\}_y - \{\zeta \psi_y\}_x \approx D_{0y}(\zeta D_{0x}\psi) - D_{0x}(\zeta D_{0y}\psi)$$

Neither of these conserves energy or enstrophy, but it was shown by Arakawa that the average of the three does. Salmon & Talley (1989) showed very elegantly how this property can be generalized to other discretizations such as finite elements or spectral methods.

Unfortunately, this approach does not simply carry over to the SWE, as the flow field usually is divergent and there is an additional variable h to be determined. Several attempts have been made to design quadartically conservative methods for the SWE. There does not seem to be any which conserves both linear and quadratic properties. In most cases, the derivation starts from the nonconservative momentum equations written as:

$$\frac{\partial u}{\partial t} - qav + \frac{\partial}{\partial x}\{ga + \tfrac{1}{2}(u^2 + v^2)\} = 0 \tag{7.40a}$$

$$\frac{\partial v}{\partial t} + qau + \frac{\partial}{\partial y}\{ga + \tfrac{1}{2}(u^2 + v^2)\} = 0 \tag{7.40b}$$

where q is the potential vorticity defined in section 3.3.4. and where all frictional terms are disregarded. Sadourny (1975) gives the following (potential) enstrophy-conserving (but not energy-conserving) semi-discretization:

$$\frac{\partial u}{\partial t} - M_y\{q\ M_x(vM_ya)\} + D_{0x}\{ga + \tfrac{1}{2}(M_xu^2 + M_yv^2)\} = 0 \tag{7.41}$$

and similarly for the other equation. The continuity equation is identical to (6.5). Here, q is defined in the C-grid (fig. 6.2) in the center of a box:

$$q_{i,j+1} = \frac{D_{0x}v - D_{0y}u + f}{M_xM_ya}$$

A similar scheme which does conserve energy but not enstrophy is also given by Sadourny (not reproduced here). For actual application, you will have to discretise time as well, which is discussed in chapter 8. The consequence is that exact conservation is no longer possible except in a few special cases. Fig. 7.8a, produced by Sadourny using a leap-frog time discretization with some small diffusion built in, shows this: even in an energy-conserving method is the energy not exactly constant. The interesting thing is that enstrophy can grow, though not unlimited: the result is that energy remains finite but gets piled up at small wave numbers. This happens the more so if the grid is finer. Fig. 7.8b shows the corresponding result for the enstrophy-conserving scheme given above: energy is again not exactly constant, but as enstrophy cannot grow (it is even slightly damped by viscosity), energy apparently is not spuriously transferred to small scales.

Arakawa and Lamb (1981) have derived a similar method which conserves energy and potential enstrophy for the shallow-water equations. However, the aim is met only by allowing some artificial numerical terms in the equations which are of second order in the grid size. The derivation is quite involved and is not reproduced here. Arakawa and Hsu

(1990) give a variant of this method in which enstrophy is not constant but slightly dissipated by a diffusion term proposed by Sadourny & Basdevant (1985). Janjic (1984) discusses other forms, both on C and E grids.

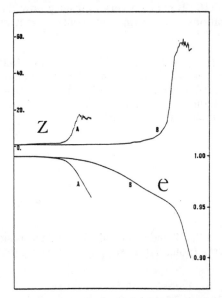 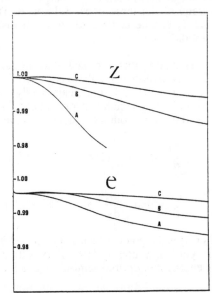

Fig. 7.8. Experiments by Sadourny (1975). Total energy E and total enstrophy Z are shown as functions of time.
a. Energy-conserving method (A = coarse grid, B = fine grid)
b. Enstrophy-conserving method (viscosity decreases from A to C)

Chapter 8

Time integration methods

8.1. Systems of ordinary differential equations

The process of semi-discretization discussed in chapter 6 produces a system of ordinary differential equations with respect to time:

$$M \frac{d\mathbf{v}}{dt} + A \mathbf{v} = 0 \qquad (8.1)$$

where $\mathbf{v}$ contains all unknowns and M is a "mass matrix"; it is nontrivial in finite-element methods but is unity in most other cases. The number of equations equals the number of grid-point values and may be very large. Several standard numerical methods exist to integrate such systems of differential equations. They are reviewed in sections 8.2...8.3.

As the shallow-water equations contain various physical processes (advection, viscosity, wave propagation, Coriolis acceleration), it is not obvious that such an "overall" time-discretization is optimal. Many attempts have been made to tailor the discretization to the separate processes. Different methods are used for different terms in the equations, e.g. explicit differences for advection and implicit ones for wave propagation. Various forms of such "mixed" discretizations are discussed in sections 8.4...8.7. It is then no longer possible to separate time- and space-discretization errors and you will have to consider the combined effect.There are several motivations for this mixed approach:
   * Numerical errors for the separate processes may tend to cancel.
   * Stability restrictions may be removed by treating the "fast" processes
      implicitly.
   * Subprocesses may even be run at different time steps for efficiency.

8.2. Explicit methods

8.2.1. Leap-frog

The variables are discretized at time levels $n\Delta t$, indicated by a superscript n (this is not an exponent). A popular method is the leap-frog method:

$$M (\mathbf{v}^{n+1} - \mathbf{v}^{n-1}) + 2 \Delta t A \mathbf{v}^n = 0 \qquad (8.2)$$

It is explicit only if the mass matrix M is diagonal. In the FEM this requires mass-lumping (e.g. Praagman, 1979). You need two starting levels, only one of which is provided by the initial conditions. You will have to use another method to do the first time step. The truncation error due to time discretization is of the order Δt^2. This is relatively favorable considering the simplicity of the method, which explains its popularity.

Stability of the method is investigated using Von Neumann's method. You can write the numerical solution as a linear combination of the eigenvectors v_i (corresponding to eigenvalues $-\mu_i$) of the matrix $M^{-1}A$:

$$v^n = \sum_i a_i v_i$$

Then

$$v^{n+1} = \sum_i \rho_i a_i v_i$$

where ρ_i is the (generally complex) amplification factor. The solution remains bounded if for each component $|\rho| \leq 1$. For the leap-frog case, you find by substitution (suppressing the subscript i)

$$\rho^{n+1} - \rho^{n-1} - 2\mu \, \Delta t \, \rho^n = 0$$

where n is an exponent. For each eigenvalue, there are two roots

$$\rho = \mu \, \Delta t \pm \sqrt{1 + (\mu \, \Delta t)^2} \tag{8.3}$$

If μ has a nonzero real part, you can verify that one of the roots will be outside the unit circle. Stability is therefore obtained only if all eigenvalues μ are purely imaginary and

$$|\mu \, \Delta t| \leq 1 \tag{8.4}$$

In some cases the eigenvalues are indeed imaginary (gravity waves, vorticity waves, Rossby waves) but not if any friction occurs. This can be fixed by a modification of the method; such modifications are discussed in section 8.4.2.

One of the roots (with the minus sign) in (8.3) is spurious; if $\Delta t \to 0$ it approaches -1 which indicates an oscillation in time. The phenomenon is called time splitting; it is not an instability as the amplitude does not grow in time. However, the oscillations can be quite unpleasant; they can be suppressed by occasionally (say, every 20 or 50 time steps) averaging two successive time levels of the solution.

If the boundary conditions are periodic, you can show that the eigenvectors are harmonic waves, i.e. composed of grid-point values

$$(u,v,h)^n_{m,j} = (u_0, v_0, h_0) \; exp\{i(k_1 x_m + k_2 y_j)\}$$

and the eigenvalues are $\mu = -i\nu$ as determined in ch. 7 for the various wave types and spatial discretizations. This case is considered in the Von Neumann stability analysis. The analysis is therefore not completely general and in particular ignores the effect of other types of boundary conditions. For that effect, see ch. 10.

For gravity waves on grid A ((7.6) with Rossby radius $R \to \infty$) for example, the eigenvalues are purely imaginary and the stability condition (8.4) gives

$$\sqrt{ga}\,\frac{\Delta t}{\Delta x}\,(sin^2\,\xi + sin^2\,\eta)^{\frac{1}{2}} \leq 1$$

which should be satisfied for any value of $\xi = k_1\Delta x$, $\eta = k_2\Delta y$. The worst case occurs for $\xi = \eta = \pi/2$:

$$\sigma = \sqrt{ga}\,\frac{\Delta t}{\Delta x} \leq 1/\sqrt{2} \qquad\qquad (8.5)$$

This is the *Courant-Friedrichs-Lewy* (CFL) condition which says that the time step should not be larger than the travel time of the wave over one grid interval. The quantity σ is known as the Courant number. Eq. (8.5) says that for a given grid-size you cannot take the time step arbitrarily large. In 1-d gravity-wave propagation, you find $\sigma \leq 1$, so the 2-d case is more restrictive by a factor $\sqrt{2}$. The reason is that a wave travelling at an angle $\pi/2$ relative to the grid "sees" grid points at a smallest distance $\Delta x/\sqrt{2}$ and the travel time is $\Delta x/\sqrt{2ga}$. There are ways to avoid this restriction as discussed in section 8.4.

8.2.2. Forward Euler method

A very simple explicit one-step method is the forward-Euler method

$$M\,(v^{n+1} - v^n) + \Delta t\,A\,v^n = 0 \qquad\qquad (8.6)$$

The amplification factor is

$$\rho = 1 + \mu\,\Delta t \qquad\qquad (8.7)$$

Stability is obtained if $\mu\Delta t$ is within the circle indicated in fig. 8.1. This excludes cases with purely imaginary μ and the method is therefore not suitable for general purposes. Variants with better stability properties can be found in section 8.4.1.

8.2.3. Runge Kutta

More suitable methods to solve systems of ODE's is the Runge-Kutta methods. The most popular form is the 4th order RK4 method, applied, e.g., by Praagman (1979) and Wubs (1987). For the case where $M = I$, the 4-stage procedure is

$$v^{n+1} = v^n - \frac{1}{6}\,(K_1 + 2K_2 + 2K_3 + K_4) \qquad\qquad (8.8)$$

$$\mathbf{K}_1 = \Delta t\, A\, \mathbf{v}^n$$

$$\mathbf{K}_2 = \Delta t\, A\, (\mathbf{v}^n - \tfrac{1}{2}\mathbf{K}_1)$$

$$\mathbf{K}_3 = \Delta t\, A\, (\mathbf{v}^n - \tfrac{1}{2}\mathbf{K}_2)$$

$$\mathbf{K}_4 = \Delta t\, A\, (\mathbf{v}^n - \mathbf{K}_3)$$

The amplification factor can be found for the constant-coefficient case:

$$\rho = 1 + \mu\,\Delta t + \tfrac{1}{2}(\mu\,\Delta t)^2 + \tfrac{1}{6}(\mu\,\Delta t)^3 + \tfrac{1}{24}(\mu\,\Delta t)^4 \tag{8.9}$$

The stability region is illustrated in fig. 8.1. For purely imaginary eigenvalues (undamped waves), the allowable time step is about 3 times larger than for leap-frog; for real eigenvalues (damped waves), about 1.5 times larger than forward Euler. However, for a complete time step the amount of work is 4 times that of a "simple" method. As you will see in ch. 9, this is more than compensated by the better accuracy of the RK4 method, which allows you to take larger time steps.

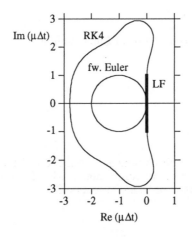

Fig. 8.1. Stability regions in the complex $\mu\Delta t$ plane for the leap-frog, forward Euler and 4-th order Runge Kutta methods

8.3. Implicit methods

All explicit methods suffer from some form of the CFL condition, which requires relatively small time steps to be taken even if gravity waves are not interesting from a physical point of view, e.g. if you are trying to compute a steady state or slowly moving flood or Rossby waves. As you are using the complete system of equations, the gravity wave just happens to be present and you cannot violate the CFL condition on the penalty of instability. To avoid such restrictions, implicit methods can be used such as the Crank-Nicolson method

$$M (v^{n+1} - v^n) + \Delta t \, A \, \{(\theta v^n + (1 - \theta)v^{n+1}\} = 0 \tag{8.10}$$

where θ is an adjustable parameter. The method is sometimes called the Crank-Nicolson method. It has second order accuracy in time if the weighting coefficient is chosen as $\theta = 1/2$. A large system of (generally nonlinear) algebraic equations results which has to be solved at each time step. Whether or not there is a mass matrix M does not make a difference for the computational complexity. The system can be split into a number of smaller systems by the ADI method discussed in section 8.5. Wilders et al (1988) have proposed the fully implicit method (8.10) in order to avoid certain accuracy problems connected with the ADI method.

The great advantage of (8.10) is its unconditional stability obtained if $\theta \geq 1/2$ is chosen. The amplification factor

$$\rho = \frac{1 + (1 - \theta) \, \mu \, \Delta t}{1 - \theta \, \mu \, \Delta t} \tag{8.11}$$

will then be within the unit circle for any value of the Courant number.

8.4. Semi-implicit methods

8.4.1. Semi-implicit Euler

The forward Euler method (section 8.2.2) gets unstable if there are eigenvalues on the imaginary axis, which is usually the case due to the presence of gravity waves. Implicit methods do not have this problem, so you could try to stabilize the method by treating some of the terms implicitly. This can be done, e.g., by using the values on the new time level $n+1$ in subsequent equations as soon as they have been computed. The method then remains effectively explicit in the sense that you do not need to solve systems of algebraic equations at each time step (Fischer, 1959, Sielecki, 1968, Van der Houwen 1977). Several variants are possible with similar properties. Here only one is discussed. On grid E the difference equations are chosen as

$$\frac{u^{n+1} - u^n}{\Delta t} + U \, M_y D_{0x} u^n - f \, v^n + g \, D_{0x} h^n + r \, u^n = 0$$

$$\frac{v^{n+1} - v^n}{\Delta t} + U \, M_y D_{0x} v^n + f \, u^{n+1} + g \, D_{0y} h^n + r \, v^n = 0 \tag{8.12}$$

$$\frac{h^{n+1} - h^n}{\Delta t} + U \, M_y D_{0x} h^n + a(\, D_{0x} u^{n+1} + D_{0y} \, v^{n+1}) = 0$$

Note carefully how the Coriolis and friction terms are treated. This is the linearized form where advection terms in y-direction have been omitted for simplicity. It is straightforwardly generalized to the nonlinear case. The method is also known as the forward-backward method. A general stability analysis is quite involved (Van der Houwen 1977). For the special case where $U=V=0$ and $r=0$, you find a cubic equation for the amplification factor:

$$(\rho-1) \{ \rho^2 - (2 - \sigma_x^2 \sin^2\xi - \sigma_y^2 \sin^2\eta + f \Delta t \, \sigma_x\sigma_y \sin \xi \sin \eta - f^2 \Delta t^2) \rho + 1 \} = 0$$

The Courant numbers in x,y directions are $\sigma_x = \sqrt{ga} \, \Delta t/\Delta x$, $\sigma_y = \sqrt{ga} \, \Delta t/\Delta y$. There is one root $\rho = 1$ (corresponding to the vorticity wave). The other two roots should be complex conjugate, otherwise one of them will be outside the unit circle, their product being 1. The discriminant should therefore be negative:

$$-2 \le 2 - \sigma_x^2 \sin^2\xi - \sigma_y^2 \sin^2\eta + f \Delta t \, \sigma_x\sigma_y \sin \xi \sin \eta - f^2 \Delta t^2 \le 2 \qquad (8.13)$$

This should be valid for all possible values of $\sin \xi$ and $\sin \eta$. The quadratic expression

$$- \sigma_x^2 \sin^2\xi - \sigma_y^2 \sin^2\eta + f \Delta t \, \sigma_x\sigma_y \sin \xi \sin \eta$$

in terms of $\sin \xi$ and $\sin \eta$ is negative definite if

$$(f \Delta t \, \sigma_x\sigma_y)^2 - 4 \sigma_x^2 \sigma_y^2 \le 0 \quad \rightarrow \quad f \Delta t \le 2 \qquad (8.14)$$

This is sufficient for the right-hand side of (8.13) to be satisfied. For most applications this will not be a serious limitation as f^{-1} is approximately 3 hours. The quadratic expression then has only one extremal value at $\sin \xi =\sin \eta =0$ for which inequality (8.13) is satisfied. The expression takes its boundary extremum in the corner points: $\sin \xi =\sin \eta = \pm 1$ and the left-hand inequality of (8.13) then yields

$$\sigma_x^2 + \sigma_y^2 \pm f \Delta t \, \sigma_x\sigma_y + f^2 \Delta t^2 \le 4 \qquad (8.15)$$

This is illustrated in fig. 8.2 for $f \Delta t = 0.5$ and 1.5. The Courant numbers σ_x, σ_y should be within both ellipses. For $f = 0$ (and $\Delta x = \Delta y$) you find

$$\sigma = \sqrt{ga} \, \frac{\Delta t}{\Delta x} \le \sqrt{2} \qquad (8.16)$$

which looks very much like the standard 2-d condition (8.5) but is larger by a factor of 2. This can be explained by considering the numerical dependence region for one variable, say $h_{k,j}$ at time $n+1$ (fig. 8.3). From (8.12.c), it depends on $u_{k+1,j}$, $u_{k-1,j}$, $v_{k,j+1}$, $v_{k,j-1}$ (each at level $n+1$) and these in turn depend on values of h at the old time level n in grid points $(k+2,j)$, $(k-2,j)$, $(k,j+2)$, $(k,j-2)$. Consequently, the effective grid size during one time step is $2\Delta x$. Again the wave travelling in a direction $\pi/4$ relative to the grid is the critical one; it meets grid points at distances $\Delta x\sqrt{2}$ which gives the condition (8.16).

In a similar way, the stability can be studied for the case of bottom friction: $f= 0, r \ne 0$. Then the equation for the amplification factors becomes

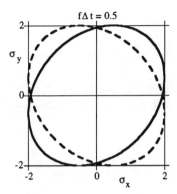

Fig. 8.2. Stability region for semi-implicit Euler (forward-backward) method. The combination σ_x, σ_y should be inside both ellipses.

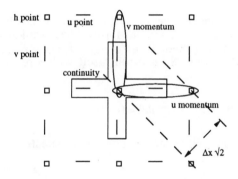

Fig. 8.3. Dependence region for semi-implicit Euler (forward-backward) method

$$(\rho - 1 + r\Delta t) \{ \rho^2 - (2 - r\Delta t - \sigma_x^2 \sin^2\xi - \sigma_y^2 \sin^2\eta) \rho + 1 - r\Delta t \} = 0$$

The "slow" root is now

$$\rho = 1 - r\Delta t$$

which is less than unity if $r\Delta t \leq 2$. With reasonable values for r, this gives a time-step limit of the order of one hour. The other roots may be complex or real. If you introduce the abbreviations

$$p = \frac{1}{2}\{\sigma_x^2 \sin^2 \xi + \sigma_y^2 \sin^2 \eta\} \qquad q = \frac{1}{2}r\Delta t$$

you can find the roots to be complex if

$$(q + p)^2 \leq 2p \tag{8.17}$$

Their modulus is then less than unity under the condition $r\Delta t \leq 2$. If the roots are real, it can easily be shown that both are less than 1. On the other hand, they should also be larger than -1, which is the case if

$$2q + p \leq 2$$

Condition (8.17) is included in this, so the stability condition for the semi-implicit Euler scheme with bottom friction becomes

$$\sigma_x^2 + \sigma_y^2 + 2 r\Delta t \leq 4 \tag{8.18}$$

The general case $f \neq 0$ and $r \neq 0$ is difficult to analyze, but you can expect to find some combination of conditions (8.15) and (8.18).

The truncation error of the semi-implicit Euler method is in general only first order in time as the terms are not centered in time, even with the forward-backward treatment. Only in the case of pure gravity-wave propagation ($U = V = 0, f = r = 0, a = $ constant) can you obtain a higher accuracy. Subtracting (8.12c) from itself with n replaced by $n-1$ and substituting (8.12a,b), you get

$$h^{n+1} - 2h^n + h^{n-1} - ga\,\Delta t^2\,(D_{0x}^2 + D_{0y}^2)h^n = 0$$

which is a second-order (because fully time-centered) approximation of the 2-d wave equation (4.5).

8.4.2. Semi-implicit leap-frog

The standard leap-frog method (section 8.2.1) is unstable if the system has eigenvalues with a nonzero real part. This is the case if bottom-friction terms are included. A remedy is to treat those terms implicitly (Holsters, 1961, Platzmann, 1972, Van der Houwen 1977). For example, on grid E, (6.6b) can be discretized as (again with advection only in x-direction for simplicity)

$$\frac{u^{n+1} - u^{n-1}}{2\Delta t} + U M_y D_{0x}u^n - f v^n + g D_{0x}h^n + \frac{1}{2} r (u^{n+1} + u^{n-1}) = 0 \tag{8.19}$$

The occurrence of u^{n+1} in the frictional term does not affect the explicit solution as it remains the only unknown in the equation. The averaging does not disturb the (second) *order* of the approximation. In the actual nonlinear equations, the bottom friction terms could be approximated in a number of ways:

$$W_x = \frac{c_f}{a} u \sqrt{u^2 + v^2} \approx \frac{1}{2} (W_x^{n-1} + W_x^{n+1}) \approx u^{n+1} \left(\frac{c_f}{a}\sqrt{u^2 + v^2}\right)^{n-1}$$

the former being the more correct one but it results in a coupling between u,v and a at time level $n+1$. The latter method is the simpler one as it produces linear and uncoupled equations at time level $n+1$. If $r = 0$, you have the normal leap-frog method (see section 8.2.1). Returning to the linearized case, if $r \neq 0$, the amplification factors follow from the equation

$$\{\rho^2 - 1 + r\Delta t \,(\rho^2 + 1)\}[\{\rho^2 - 1 + r\Delta t \,(\rho^2 + 1)\}(\rho^2 - 1) +$$
$$+ 4\,\rho^2(\sigma_x^2 sin^2\xi + \sigma_y^2 sin^2\eta)] = 0$$

Note that this is of double order compared with the previous section, which indicates that there is a spurious root connected to each physical root. This is typical for methods of the leap-frog variety. For the slow waves, you find

$$\rho^2 = \frac{1 - r\Delta t}{1 + r\Delta t}$$

which is stable if $r\Delta t \leq 1$. To find the other roots, introduce the same abbreviations as in the previous section together with $\rho^2 = x$. This gives

$$(1 + 2q)\, x^2 - 2\,(1 - p)\, x + 1 - 2\,q = 0$$

If the roots are complex, they are again less than unity if $r\Delta t \leq 1$. This occurs if

$$(1 - p)^2 \leq 1 - 4q^2$$

On the other hand, if the roots are real, it is again simple to show that they are less than 1; to have them greater than -1, you find $p \leq 2$. This includes again the complex case, so you find as the stability condition just the standard one for leap-frog which agrees with (8.5):

$$\sigma_x^2 + \sigma_y^2 \leq 1 \tag{8.20}$$

There are other ways to treat terms implicitly. In meteorology, one is not very much interested in gravity waves and wants to have the time step limited only by the relevant slow waves. The same is true if you are trying to approach a steady flow. This can be accomplished by taking the wave-propagation part implicitly. Various authors have proposed this (Kwizak & Roberts, 1971, Elvius & Sundström, 1973 and in a finite-element fashion Niemeyer, 1979). On the C grid, this gives from (6.8)

$$\frac{u^{n+1} - u^{n-1}}{2\Delta t} + U\,M_x D_{0x} u^n - f\,M_x M_y v^n + \frac{1}{2} g\,D_{0x}(h^{n-1} + h^{n+1}) + r\,u^n = 0$$

$$\frac{v^{n+1} - v^{n-1}}{2\Delta t} + U\,M_x D_{0x} v^n + f\,M_x M_y u^n + \tfrac{1}{2} g\,D_{0y}(h^{n-1} + h^{n+1}) + r\,v^n = 0$$

$$\tag{8.21}$$

$$\frac{h^{n+1} - h^{n-1}}{2\Delta t} + U\,M_x D_{0x} h^n + \tfrac{1}{2} a\{\, D_{0x}(u^{n+1} + u^{n-1}) + D_{0y}\,(v^{n+1} + v^{n-1})\} = 0$$

The result is now an implicit method as the unknowns in all grid points are coupled. You can eliminate the velocity components at level $n+1$ and obtain a discrete Helmholtz- type equation for the water-level h^{n+1}:

$$h^{n+1} - \tfrac{1}{4} ga\,\Delta t^2 (D_{0x}^2 + D_{0y}^2)\,h^{n+1} = R_3 - \tfrac{1}{2}\,a\,\Delta t\,(D_{0x} R_1 + D_{0y} R_2) \tag{8.22}$$

where R_i are the terms on time level n brought to the right-hand side in (8.21). Eq. (8.22) has to be solved at each time step. The coefficients on the left-hand side depend only on the water depth which often does not vary too much with time. If you would substitute the constant (mean) part of the depth, the coefficient matrix would be constant. You would have to treat *variations* of the depth in (6.8a) explicitly (this was proposed by Kinnmark, 1986). The system (8.22) moreover has a simple structure for which efficient numerical techniques are available. An LU splitting could be computed once and for all and used in each time step again. This idea can also be combined with an ADI splitting (Elvius & Sundström, 1973).

The time differencing is such that second order accuracy is retained. The stability of the method is now determined by the slow-wave speed. Elvius & Sundström do not include Coriolis terms in their stability analysis. If you do, but neglect bottom friction terms, you find a quartic equation for the gravity-wave amplification factor ($\Delta x = \Delta y$ for simplicity):

$$(1 + 2\,\sigma^2 sin^2 \xi)\,\rho^4 - \{2 - 4\,\sigma^2 sin^2 \xi - 4\,(f\,\Delta t)^2 cos^4 \xi\}\rho^2 + 1 + 2\,\sigma^2 sin^2 \xi = 0$$

Two roots correspond to physical solutions; the other two are spurious as usual in leap-frog type methods. You find either two real roots for ρ^2, or two complex conjugate ones. In the former case, one of the roots will exceed unity as the product of the roots is 1, and the method is unstable. In the latter case, both roots are on the unit circle. You have stability only in this case, obtained if the discriminant is negative:

or

$$1 - 2\,\sigma^2 sin^2 \xi - 2\,(f\,\Delta t)^2 cos^4 \xi \ge -1 - 2\,\sigma^2 sin^2 \xi$$

$$f\,\Delta t \le 1 \tag{8.23}$$

The time step can therefore be taken an order of magnitude larger than for the effectively explicit method (8.19), at least from a stability point of view. This offsets the additional effort needed to solve (8.22) at each time step. Whether this remains true in view of accuracy is discussed in chapter 9.

8.5. ADI methods

Fully implicit methods, mentioned in section 8.3, have the great disadvantage that a large system of algebraic equations has to be solved at each time step, which may be quite a job even if you locally linearize them. A very popular method to avoid this while still retaining the major advantages is the alternating-direction implicit (ADI) method. Its origin is in papers by Peaceman & Rachford (1955), Douglas & Rachford (1958), Douglas & Gunn (1964). It has been used extensively in compressible aerodynamics, e.g. by Beam & Warming (1976,1978), Briley & McDonald (1980). For the SWE, several ADI type methods have been proposed by, e.g. Leendertse (1967), Gustafsson (1971), Abbott et al (1972), Fairweather & Navon (1980), Stelling (1983).

The system of equations originating from an implicit (Crank-Nicolson) discretization of (4.3) can be written as

$$(I + \theta P + \theta Q)\, \mathbf{v}^{n+1} = \{I - (1-\theta)\,(P+Q)\}\, \mathbf{v}^n \tag{8.24}$$

where P and Q are spatial discretizations of

$$P = \Delta t\left(A\frac{\partial}{\partial x} + C_1\right) \qquad Q = \Delta t\left(B\frac{\partial}{\partial y} + C_2\right)$$

with some splitting of the non-derivative C term containing bottom friction and Coriolis terms. For example, on the A grid, a splitting proposed by Gustafsson (1971) is

$$P = \Delta t\begin{pmatrix} U\,D_{0x}+ r & 0 & g\,D_{0x} \\ f & U\,D_{0x} & 0 \\ a\,D_{0x} & 0 & U\,D_{0x} \end{pmatrix}$$

$$Q = \Delta t\begin{pmatrix} V\,D_{0y} & -f & 0 \\ 0 & V\,D_{0y}+ r & g\,D_{0y} \\ 0 & a\,D_{0y} & V\,D_{0y} \end{pmatrix}$$

The truncation error of (8.24) is first order in time, unless $\theta = 0.5$, in which case it is second order. In the "delta formulation" (Beam & Warming), (8.24) becomes

$$(I + \theta P + \theta Q)\,\Delta\mathbf{v}^{n+1} = -(P+Q)\,\mathbf{v}^n \qquad \Delta\mathbf{v}^{n+1} = \mathbf{v}^{n+1} - \mathbf{v}^n$$

The left-hand side can be factored approximately:

$$(I + \theta P)(I + \theta Q)\,\Delta\mathbf{v}^{n+1} = -(P+Q)\,\mathbf{v}^n$$

which introduces an error of $O(\Delta t^3)$ as $\Delta\mathbf{v} = O(\Delta t)$. However, the time discretization did already include an error of this order so the order of the approximation is not changed (although the *magnitude* of the error might be affected). Introducing auxiliary values $\Delta\mathbf{v}^*$, you get

$$(I + \theta P) \, \Delta\mathbf{v}^* = - (P+Q) \, \mathbf{v}^n$$

$$(I + \theta Q) \, \Delta\mathbf{v}^{n+1} = \Delta\mathbf{v}^*$$

(8.25)

If written in the original variables instead of the delta form, this is the method as given by Douglas & Gunn (1964) and Douglas & Rachford (1958). Each of these systems of equations involves derivatives (or rather differences) in one direction only and they are solved alternatingly, which explains the name of the method. The advantage is that implicit coupling of variables occurs only along lines of constant x or y, without coupling in the other direction. Eq.(8.25a) therefore consists of a number of uncoupled systems of equations for each row of grid points separately with a narrow band-width; (8.25b) does the same for columns. Solving these relatively small systems is more efficient in computer time and memory than solving the single large system of (8.24). Moreover, the small systems for each row or column can very well be solved in parallel.

Another formulation exists, which can be obtained by defining half time-step values as

$$\mathbf{v}^{n+1/2} = \tfrac{1}{2} (\mathbf{v}^n + \mathbf{v}^*) = \mathbf{v}^n + \tfrac{1}{2}\Delta\mathbf{v}^*$$

Introducing this into (8.25), taking $\theta = 0.5$ and adding the two equations gives a nicely symmetric form which shows the ADI character more clearly:

$$(I + \tfrac{1}{2} P) \, \mathbf{v}^{n+1/2} = (I - \tfrac{1}{2} Q) \, \mathbf{v}^n$$

$$(I + \tfrac{1}{2} Q) \, \mathbf{v}^{n+1} = (I - \tfrac{1}{2} P) \, \mathbf{v}^{n+1/2}$$

(8.26)

This is the form introduced by Peaceman & Rachford. It just treats the x-derivatives implicitly and they-derivatives explicitly during half a time step, and conversely in the second half step. One advantage is that $\mathbf{v}^n$ does not occur any longer as soon as you have computed $\mathbf{v}^{n+1/2}$; this saves computer memory. It does not seem possible to get a similarly simple form for arbitrary θ. Leendertse (1967) used this approach on the C grid with additional modifications in the nonlinear terms to obtain scalar tridiagonal equations. In linear form, this would amount to the system

$$\frac{u^{n+1/2} - u^n}{\Delta t/2} + (U \, M_x D_{0x} + r) \, u^n + g \, D_{0x} h^{n+1/2} - f \, M_x M_y v^n = 0$$

$$\frac{v^{n+1/2} - v^n}{\Delta t/2} + U \, M_x D_{0x} v^n + g \, D_{0x} h^n + f \, M_x M_y u^n + r \, v^{n+1/2} = 0$$

$$\frac{h^{n+1/2} - h^n}{\Delta t/2} + U \, M_x D_{0x} h^n + a \, D_{0x} u^{n+1/2} + a \, D_{0y} v^n = 0$$

(8.27)

$$\frac{u^{n+1} - u^{n+1/2}}{\Delta t/2} + U \, M_x D_{0x} \, u^{n+1/2} + g \, D_{0x} h^{n+1/2} - f \, M_x M_y v^{n+1} + r \, u^{n+1} = 0$$

$$\frac{v^{n+1} - v^{n+1/2}}{\Delta t/2} + U M_x D_{0x} v^{n+1/2} + g D_{0x} h^{n+1} + f M_x M_y u^{n+1/2} + r\, v^{n+1/2} = 0$$

$$\frac{h^{n+1} - h^{n+1/2}}{\Delta t/2} + U M_x D_{0x} h^{n+1/2} + a D_{0x} u^{n+1/2} + a D_{0y} v^{n+1} = 0$$

Note that (8.27b and d) are effectively explicit once the other variables have been solved. The advective terms are lagged behind in all equations and are therefore not properly time-centered. This is not harmful for cases where advection is unimportant. In nonlinear form, the advective terms are treated as (e.g. in (8.27a))

$$u^{n+1/2} M_x D_{0x} u^n$$

which does not make the equations nonlinear but still is not properly centered. Similarly, in the nonlinear frictional terms, approximations are used of the form (e.g. in (8.27b))

$$\frac{v}{a}\sqrt{u^2 + v^2} \approx \frac{v^{n+1/2}}{M_y a^{n+1/2}}\sqrt{(M_x M_y u^{n+1/2})^2 + (v^n)^2}$$

in which only $v^{n+1/2}$ is unknown because the other variables have already been computed from the tridiagonal systems originating from (8.27a and c). Averaging operators are included because of the C grid.

Stelling (1983) introduced a great number of improvements to restore the accuracy of the method. In the present linearized discussion, this would only affect the treatment of the advective terms (section 11.5) and the order in which bottom friction and Coriolis terms would be taken into account. For further details see the original reference.

You notice that both forms (8.25) and (8.26) of the method are equivalent, at least for linear equations. This is no longer true for nonlinear equations. In that case, the coefficients P and Q will be unknown, depending on the new values of the variables. Several methods have been proposed to solve this. Gustafsson (1971) used a straightforward iteration method (either substitution or a quasi Newton method). Fairweather & Navon (1980) predicted the coefficients by extrapolation from previous time steps, so that linear equations result and no iteration is needed. Beam & Warming used the delta form with local linearization with respect to the known time level and solved the linearized equations only once per time step. This amounts to one Newton iteration and is formally sufficient to get second-order time accuracy. Leendertse (1967) and Stelling (1983) used several ad-hoc approximations of the nonlinear terms in order to get linear tridiagonal systems. Iteration is needed to obtain second-order time accuracy.

Stability of the ADI method is not easy to establish. For the pure gravity-wave case (i.e. no advection, Coriolis or bottom friction), Vreugdenhil (1989) demonstrated that the method is unconditionally stable if you choose the coefficient $\theta \geq 0.5$, as in its non-split counterpart. General experience is that this remains valid if you include the other terms, but a proof of this is unknown. The difficulty with the analysis is that in general the matrices A, B and C do not have common eigenvectors.

8.6. Fractional-step methods

In fractional-step methods, several parts of the equations are treated with different numerical methods. ADI methods can be labeled fractional-step methods in this sense, but here the term is reserved for the case where the distinction is between different *physical* processes (not coordinate directions as in ADI): advection, Coriolis acceleration, wave propagation, friction. An approach in this sense was proposed by Gadd (1978); it consists of three steps:

(i)   "Adjustment stage" in which wave propagation and Coriolis are treated using a semi-implicit (forward-backward) method as discussed in section 8.4.1. A number of M time steps of magnitude δt (each smaller than the global time step) are taken.

(ii)   Advection (and diffusion if present) with a larger time step $\Delta t = M \delta t$, solved with a Lax-Wendroff type method which is second order accurate in time

(iii)   Friction, which is solved exactly for a quadratic friction law in 2-d during a time step Δt.

A similar idea was put forward by Benqué et al (1982) and this form is discussed as a typical example. The Arakawa E-grid is used and the splitting in physical processes is as follows.

Advection is solved by means of the method of characteristics. From each grid point, a characteristic (stream line) is traced back from the new time level $n+1$ to the previous one n. There, the variables are interpolated from surrounding grid points using higher-order interpolation. The interpolated values are then supposed to be conserved along the stream line and therefore to apply also in the grid point from which the construction started. In the present linearized form with constant advection velocity U, this would mean

$$u^{n+1/3} = u^n(x - U \, \Delta t) \tag{8.28}$$

and similarly for v and h, where $n+1/3$ is just an indication of an intermediate, physically not meaningful, level. Here, it has been assumed that interpolation is so accurate that essentially exact values are obtained on the right-hand side.

In the second step, diffusion (viscosity) is treated implicitly, together with explicit Coriolis terms; the resulting Poisson-type equations are solved iteratively by a standard ADI method. In the present analysis, diffusion is not included so you are left with

$$u^{n+2/3} = u^{n+1/3} + f \, \Delta t v^{n+1/3} \tag{8.29}$$

and similarly for v.

The third ("propagation") step is similar to Gadd's adjustment step: it includes gravity and velocity divergence, but here bottom friction is included:

$$\frac{u^{n+1} - u^{n+2/3}}{\Delta t} + \frac{1}{2}\left(g \, D_{0x}(h^{n+1} + h^n) + r\,(u^{n+1} + u^{n+2/3})\right) = 0 \tag{8.30}$$

(similarly for v)

$$\frac{h^{n+1} - h^{n+1/3}}{\Delta t} + \frac{1}{2} a \left(D_{0x}(u^{n+1} + u^n) + D_{0y} (v^{n+1} + v^n) \right) = 0 \qquad (8.31)$$

The set (8.30) (8.31) is implicit and solved iteratively by an ADI method after elimination of (u,v); this is similar to (8.21) (Elvius & Sundström, 1973).

To get an impression of the numerical accuracy, you might eliminate the intermediate levels, which can be done for the linearized equations; this gives

$$\frac{u^{n+1} - u^n(x^*)}{\Delta t} - f v^n(x^*) + \frac{1}{2} \left(g D_{0x}(h^{n+1} + h^n) + r (u^{n+1} + u^n(x^*)) + rf \Delta t\, v^n(x^*) \right) = 0$$

$$\frac{v^{n+1} - v^n(x^*)}{\Delta t} + f u^n(x^*) + \frac{1}{2} \left(g D_{0y}(h^{n+1} + h^n) + r (v^{n+1} + v^n(x^*)) - rf \Delta t\, u^n(x^*) \right) = 0$$

$$\qquad (8.32)$$

$$\frac{h^{n+1} - h^n(x^*)}{\Delta t} + \frac{1}{2} a \left(D_{0x}(u^{n+1} + u^n) + D_{0y} (v^{n+1} + v^n) \right) = 0$$

with the notation $x^* = x - U\Delta t$ (similarly for the y direction if an advection velocity is present in that direction). This form will be used for a quantitative comparison with other methods in terms of wave propagation (chapter 9). Formal accuracy in time is only $O(\Delta t)$ due to fractional time-stepping; you can see this, e.g., in the last terms of (8.32a,b). This could be increased to second-order accuracy by interchanging the order of substeps during the next time step (Strang, 1968). A more formal method of splitting, which allows higher-order accuracy, has been described by Maday et al (1990)

8.7. Riemann solvers

A quite different idea of splitting is based on the existence of distinct wave types in the solution of the SWE. In compressible aerodynamics, an important part is played by shock waves (see also section 3.4). A standard shock-wave problem describing the interaction of two different constant-property (velocity, pressure) states is called a Riemann problem. Methods in which a numerical solution is built up from a series of elementary Riemann problems at cell interfaces are called Riemann solvers. You can find reviews of such methods in (e.g.) LeVeque (1990) or Hirsch (vol.2, 1990), which are the basis for the following discussion. For applications to the SWE, see Tan (1992), Priestley (1987), Glaister (1993). Many aspects which are important in aerodynamics are less so for the SWE. For example, shock waves are very uncommon in shallow-water. Therefore, only one or two typical methods are considered here to get an impression of the applicability for smooth problems. The discussion starts from 1-d linear systems with constant coefficients.

8.7.1. Linear problem in one spatial dimension

The theory of characteristics, discussed in section 4.3, can be cast in a different form for a linear system of hyperbolic equations (such as the SWE) in one spatial dimension

$$\frac{\partial \mathbf{v}}{\partial t} + A \frac{\partial \mathbf{v}}{\partial x} = 0 \tag{8.33}$$

If the matrix A has distinct real eigenvalues (characteristic speeds) and a complete set of eigenvectors (which is the condition for being strictly hyperbolic), there is a matrix R which transforms A to diagonal form:

$$A = R \Lambda R^{-1}$$

The columns of R are the eigenvectors of A; the entries of the diagonal matrix Λ are the eigenvalues c of A. Eq. (8.33) can then be written as

$$R^{-1} \frac{\partial \mathbf{v}}{\partial t} + R^{-1} A R R^{-1} \frac{\partial \mathbf{v}}{\partial x} = 0$$

so defining the characteristic variables

$$\mathbf{w} = R^{-1} \mathbf{v} \tag{8.34}$$

you get a system of decoupled equations

$$\frac{\partial \mathbf{w}}{\partial t} + \Lambda \frac{\partial \mathbf{w}}{\partial x} = 0 \tag{8.35}$$

The attractive feature of this system is that it consists of three scalar simple-wave equations (or advection equations) with the corresponding wave speed c as advection velocity. For the k-th scalar component (Riemann invariant) w_k, there is an exact solution:

$$w_k(x,t) = w_k(x - c_k \Delta t, t - \Delta t) \tag{8.36}$$

through which you can work back from any (x,t) to a boundary or initial condition. Using (8.34) you can then construct an exact solution to the original equations. Numerically, you could use, e.g., a simple first-order upwind method for each scalar equation (j indicates the grid point, n the time level):

$$(w_k)_j^{n+1} = (w_k)_j^n - c_k \Delta t \, D_{-x} w_k^n \qquad \text{(if } c_k > 0\text{)}$$

$$\tag{8.37}$$

$$(w_k)_j^{n+1} = (w_k)_j^n - c_k \Delta t \, D_{+x} w_k^n \qquad \text{(if } c_k < 0\text{)}$$

where D_{+x} and D_{-x} are forward and backward differences, respectively. The direction in which the differences are taken differs for the various components, depending on the sign of the propagation speed c_k. The method is only first-order accurate both in Δx and Δt, but it can be relatively accurate if time- and space discretisation errors cancel, which is the case if the Courant number $\sigma_k = c_k \Delta t / \Delta x$ is near unity. Actually, if it is unity, you get an exact solution, as you can see from a comparison of (8.36) and (8.37). Unfortunately, it is hard to exploit this for two reasons. Firstly, even for one characteristic component the wave speed c_k (which could be, e.g. $u + \sqrt{ga}$) is usually not constant but (slowly) varying in space and time. Secondly, there may be an order of magnitude between the various characteristic speeds. Using one set of Δx and Δt, it is impossible to get all Courant numbers near unity. The largest one must be less than unity (for stability), but you can choose it near unity. Then the smaller wave speeds will lead to small Courant numbers, accompanied by considerable inaccuracy.

The first-order upwind method can be written down for the system of equations if you define coefficient matrices corresponding to right- or left-going waves separately:

$$A^+ = R \, \Lambda^+ R^{-1} \quad A^- = R \, \Lambda^- R^{-1} \quad |A| = R \, |\Lambda| \, R^{-1}$$

where Λ^+ is the part of Λ containing positive characteristic speeds, Λ^- the part containing the negative ones, and $|\Lambda|$ the matrix with all signs made positive. Then, with a similar division of $\mathbf{w}$, (8.37) becomes

$$(\mathbf{w}^+)_j^{n+1} = (\mathbf{w}^+)_j^n - \Lambda^+ \Delta t \, D_{-x}(\mathbf{w}^+)^n$$

$$(\mathbf{w}^-)_j^{n+1} = (\mathbf{w}^-)_j^n - \Lambda^- \Delta t \, D_{+x}(\mathbf{w}^-)^n$$

As $\mathbf{w}$ is just the sum of $\mathbf{w}^+$ and $\mathbf{w}^-$, this can be combined into

$$\mathbf{w}_j^{n+1} = \mathbf{w}_j^n - \Lambda^+ \Delta t \, D_{-x}\mathbf{w}^n - \Lambda^- \Delta t D_{+x}\mathbf{w}^n$$

or $\qquad \mathbf{w}_j^{n+1} = \mathbf{w}_j^n - \Lambda \, \Delta t \, D_{0x}\mathbf{w}^n + |\Lambda| \dfrac{\Delta t \, \Delta x}{2} D_{+x}D_{-x}\mathbf{w}^n \qquad$ (8.38)

You can check that the last term corresponds to a second derivative:

$$D_{+x}D_{-x}w_j = \frac{w_{j+1} - 2 \, w_j + w_{j-1}}{\Delta x^2}$$

Transforming back to the original variables,

$$\mathbf{v}_j^{n+1} = \mathbf{v}_j^n - A \, Dt \, D_{0x}\mathbf{v}^n + |A| \frac{\Delta t}{2\Delta x}\left(\mathbf{v}_{j+1}^n - 2\mathbf{v}_j^n + \mathbf{v}_{j-1}^n\right) \qquad (8.39)$$

Even though this equation does not explicitly discern between directions of propagation, it does so by means of the "switching" mechanism implicit in the last term. For most practical problems, the first-order upwind method used here is too inaccurate. Higher-order methods are discussed in section 8.7.3.

8.7.2. Nonlinear 1-d problem

If the system of equations is nonlinear, its conservative form is

$$\frac{\partial \mathbf{v}}{\partial t} + \frac{\partial \mathbf{f}(\mathbf{v})}{\partial x} = 0 \qquad (8.40)$$

The idea of splitting into elementary waves is less simple now: an arbitrary initial state may develop into a complicated pattern of smooth regions, shock waves, etc., for which a closed-form solution is hard to give. It can be done, however, for a schematized problem, which leads to *Godunovs method* as a nonlinear extension of the first-order upwind method. It supposes a piecewise constant state in each numerical cell, as shown in fig. 8.4. The value v_j^n can then be identified with the average state in that cell. For each jump at a cell face, a Riemann problem is solved: an exact nonlinear solution is constructed, including shocks, contact discontinuities and rarefaction waves. This has the form

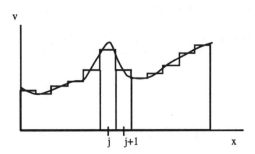

Fig. 8.4. Numerical solution built up from cell-averaged constant states.
At each cell boundary, a Riemann problem results.

$$\mathbf{v}(x - x_{j+1/2}, t - t_n) = \mathbf{v}\left(\frac{x - x_{j+1/2}}{t - t_n}, \mathbf{v}_j, \mathbf{v}_{j+1}\right) \tag{8.41}$$

i.e., the solution is constant along straight lines emanating from the original front position, and it depends of course on the two constant states on either side. One of these straight lines coincides with the cell face, so the flux across it will also be constant. By integrating (8.40) over a cell during one time step, you get a numerically conservative form (that is: the flux leaving cell j equals the flux entering cell $j+1$) which is still exact if the $\mathbf{v}$ values are cell averages:

$$\mathbf{v}_j^{n+1} = \mathbf{v}_j^n - \frac{\Delta t}{\Delta x}\left(\mathbf{f}_{j+1/2}^n - \mathbf{f}_{j-1/2}^n\right) \tag{8.42}$$

in which the constant fluxes are based on the exact values from (8.41). Next the new values at time $n+1$ are again considered as a series of constant states and the procedure is repeated. The waves from a cell face should not reach a neighbouring face within one time step, otherwise the constant value (8.41) will be corrupted. This leads to a condition

$$|c_{max}|\frac{\Delta t}{\Delta x} \le 1$$

which is recognized as the CFL condition.

Godunovs method is an exact solution of an approximate problem: the only numerical approximation consists in the schematization of an arbitrary state into a piecewise constant state. The method has some disadvantages and advantages:

(i)   For linear equations, it can be shown that the procedure is fully equivalent to the splitting into Riemann invariants described in section 8.7.1. Therefore, the accuracy for (more or less) linear waves will be no better than that of the first order upwind method. Particularly, the accuracy will be poor if the Courant number of a wave is far below unity.
(ii)   The construction of exact Riemann solutions is not simple.
(iii)   On the positive side, the method is monotone, which means that it does not add new extremes to a solution, i.e. it does not produce unphysical oscillations. Actually, it has been shown that monotone methods can be no more than 1st order accurate.

The second point can be circumvented by using an *approximate Riemann solver* instead of an exact one. Several possibilities exist to do this; an often used method is due to Roe (1981). It replaces the nonlinear Riemann problem at a cell face by a linear one, the solution of which has the form of (8.39). This turns out to be possible if you can write a flux difference in the form

$$\mathbf{f}(\mathbf{v}_1) - \mathbf{f}(\mathbf{v}_2) = \tilde{A} \ (\mathbf{v}_1 - \mathbf{v}_2)$$

where $\tilde{A}$ is the Roe matrix. Its form is discussed in section 8.7.5. Using this, (8.39) can be used locally with A replaced by $\tilde{A}$. This can also be written in a conservative form similar to (8.42):

$$\mathbf{v}_j^{n+1} = \mathbf{v}_j^n - \frac{\Delta t}{\Delta x}\left(\mathbf{f}^{*\,n}_{j+1/2} - \mathbf{f}^{*\,n}_{j-1/2}\right)$$

if the *numerical fluxes* are defined as

$$\mathbf{f}^{*}_{j+1/2} = \tfrac{1}{2}(\mathbf{f}_{j+1} + \mathbf{f}_j) - \tfrac{1}{2}|\tilde{A}| \ (\mathbf{v}_{j+1} - \mathbf{v}_j) \tag{8.43}$$

The Roe matrix should satisfy two more requirements: it should have distinct eigenvectors with real eigenvalues, and it should approach the exact Jacobian A if the two states approach one another. For each system of equations, this matrix has to be determined. For the SWE, this has been done by Priestley, 1987 (see section 8.7.5).

8.7.3. Higher order methods

The first-order upwind method is often too inaccurate, so you would want to use a more accurate method. Unfortunately, higher-order methods usually introduce unphysical oscillation ("wiggles") into the numerical solution, in particular near strong gradients or shocks. The reason (already mentioned) is that a monotone method can be no more than first-order accurate. Therefore, higher-order methods will not be monotone. However, it turns out that you can blend low- and high-order methods, such that oscillations are avoided but high-order accuracy is obtained in smooth parts. This blend may be different for the various waves, so the idea of Riemann solvers still applies.

As an example, the second-order Lax-Wendroff method is discussed, which is a notorious source of wiggles. It looks very much like (8.39) but the coefficient of the diffusion term is chosen such that second-order accuracy is obtained.

$$v^{n+1} = \mathbf{v}^n - A \; \Delta t \; D_{0x}v^n + \frac{1}{2} A^2 \, \Delta t^2 D_{+x}D_{-x}v^n \tag{8.44}$$

This can be written in the form of (8.42) with a numerical flux

$$\mathbf{f}^*_{j+1/2} = \frac{1}{2} A \, (\mathbf{v}_j + \mathbf{v}_{j+1}) - \frac{1}{2} \frac{\Delta t}{\Delta x} A^2 \, (\mathbf{v}_{j+1} - \mathbf{v}_j) \tag{8.45}$$

which can also be seen as the (low-order) upwind flux (8.43) plus a (higher-order) correction:

$$\mathbf{f}^*_{j+1/2} = \mathbf{f}^{low} + \frac{1}{2} \, (|A| - \frac{\Delta t}{\Delta x} \, A^2) \, (\mathbf{v}_{j+1} - \mathbf{v}_j) \tag{8.46}$$

You can develop the vector difference in eigenvectors $\mathbf{r}$ of A with corresponding eigenvalues c:

$$\mathbf{v}_{j+1} - \mathbf{v}_j = \sum_p \alpha_{pj} \, \mathbf{r}_p \tag{8.47}$$

Then α_{pj} is the difference in the p-th characteristic variable between grid points j and $j+1$ and

$$\mathbf{f}^*_{j+1/2} = \mathbf{f}^{low} + \frac{1}{2}\sum_p \, (|A| - \frac{\Delta t}{\Delta x} \, A^2) \, \alpha_{pj} \, \mathbf{r}_p = \mathbf{f}^{low} + \frac{1}{2}\sum_p \, (|c_p| - \frac{\Delta t}{\Delta x} \, c_p^2) \, \alpha_{pj} \, \mathbf{r}_p \tag{8.48}$$

To prevent oscillations, the higher-order correction is multiplied by a *limiter* ϕ for each eigenvector component:

$$\mathbf{f}^*_{j+1/2} = \mathbf{f}^{low} + \frac{1}{2}\sum_p \, \phi(\theta_{pj}) \, (|c_p| - \frac{\Delta t}{\Delta x} \, c_p^2) \, \alpha_{pj} \, \mathbf{r}_p \tag{8.49}$$

The function ϕ depends on the rate of change of the vector difference in any of the eigenvector directions, so it has something to do with the curvature:

$$\theta_{pj} = \frac{\alpha_{p,j-1}}{\alpha_{pj}} \; \text{if} \; c_p > 0, \; \; \theta_{pj} = \frac{\alpha_{p,j+1}}{\alpha_{pj}} \; \text{if} \; c_p < 0$$

In smooth regions, $\theta \approx 1$ and there is no great risk of spurious oscillations. Therefore, you can have the full second order accuracy using the high-order flux, so $\phi(1) = 1$. Moreover, if $\theta \le 0$, there is apparently an extremum and oscillations can occur. To prevent this, you use only the low-order flux: $\phi(\theta) = 0$ for $\theta \le 0$. Finally, it can be shown that no new oscillations will be generated if

$$0 \le \phi(\theta) \le 2 \quad \text{and} \quad 0 \le \phi(\theta) \le 2\theta$$

This gives an admissible region for the limiter, shown in fig. 8.5. Within this region, there is a freedom of choice. Several possibilities are summarized by LeVeque (1990). You will note that this scheme is nonlinear even if the original equations are linear. It will therefore not be possible to investigate the accuracy of the limiter-schemes by means of linear wave propagation, but you can assume that the accuracy will at least not be better than the full second-order Lax-Wendroff method on which it is based. Many other higher-order schemes can be substituted for the Lax-Wendrof method (see e.g. Glaister (1993) and the books by LeVeque (1990) and Hirsch (1990)).

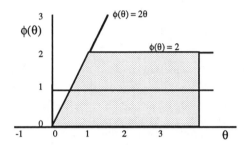

Fig. 8.5. Admissible region for limiter function

8.7.4. More dimensions

In more dimensions, it is generally not possible to bring the system of equations in diagonal form such as (8.35), for the simple reason that the matrices A, B and C in (4.3) do not have common eigenvectors.

The idea of exact Riemann solutions to a piecewise constant solution per cell becomes very complicated in more dimensions, as you will have interacting shock waves at the corners. A unique splitting in a number of waves does not exist anymore. It has been attempted to fit a predetermined number of waves to the local situation, e.g. Rumsey et al (1993). Alternatively, a fractional-step method (section 8.6) may be used consisting of 1-d Riemann solvers as sub-steps. For smooth problems this gives second-order accuracy in time if you alternate the order of the steps. For problems with shocks, Roe (1991) has shown that shocks will not be represented correctly unless they are aligned with the grid and that, generally, there is no convergence if the time step is decreased.

Most of the multi-dimensional Riemann solvers, however, just add the 1-d formulations in both directions, e.g. Alcrudo & Garcia-Navarro (1993). The Riemann formalism is applied only to the derivative part of the equations. It is not at all clear how to handle the zero-order part (Coriolis and friction terms). In the following example, simple forward Euler integration is used. Roe (1987) gives some ideas how this could be improved.

The first-order upwind formulation, applied componentwise, becomes:

$$\mathbf{v}_{k,j}^{n+1} = \mathbf{v}_{k,j}^{n} - \frac{\Delta t}{\Delta x}\left(\mathbf{f}_{k+1/2,j}^{*\,n} - \mathbf{f}_{k-1/2,j}^{*\,n}\right) - \frac{\Delta t}{\Delta y}\left(\mathbf{g}_{k,j+1/2}^{*\,n} - \mathbf{g}_{k,j-1/2}^{*\,n}\right) - \Delta t\, C\, \mathbf{v}_{k,j}^{n} \qquad (8.50)$$

where the numerical fluxes are computed by decomposition in terms of the eigenvectors (**r** and **s**) of their Jacobians A, B:

$$\mathbf{f}_{k+1/2,j}^{*} = \tfrac{1}{2}\left(\mathbf{f}_{k+1,j} + \mathbf{f}_{k,j}\right) - \tfrac{1}{2}\sum_{p} |c_p|\, \alpha_{p,k+1/2}\, \mathbf{r}_p$$

$$\mathbf{g}_{k,j+1/2}^{*} = \tfrac{1}{2}\left(\mathbf{g}_{k,j+1} + \mathbf{g}_{k,j}\right) - \tfrac{1}{2}\sum_{p} |c_p|\, \beta_{p,j+1/2}\, \mathbf{s}_p$$

Second order flux-limiting can be applied componentwise:

$$\mathbf{f}_{k+1/2,j}^{*} = \mathbf{f}^{low} + \tfrac{1}{2}\sum_{p} \phi(\theta_{p,k+1/2})\, \left(|c_p| - \frac{\Delta t}{\Delta x}\, c_p^2\right) \alpha_{p,k+1/2}\, \mathbf{r}_p \qquad (8.51)$$

and similarly for the y-direction. The wave speeds c_p are taken from the Roe matrices. However, this does *not* give second-order accuracy as the mixed derivatives in a 2-d Taylor expansion are not taken into account. Several variants of higher-order approximations with different limiters have been tested by Priestley (1987). Based on a global energy measure, he concludes that very significant errors are possible.

8.7.5. Roe decomposition for SWE

It is not a trivial matter to find a matrix satisfying the requirements given by Roe. For the 2-d SWE, the analysis has been given by Priestley (1987), see also Glaister (1993). Here, only the result is given. In the final equations, it is not the matrix that is needed but the decomposition in terms of its eigenvectors. The system of equations is used in its conservative form (2.48)...(2.50) with unknowns ga, $gau = m$, $gav = n$ (the gravitational coefficient g is introduced for convenience). Priestley (in analogy with Roe's original treatment) introduces an auxiliary vector

$$\mathbf{w} = \begin{pmatrix} \sqrt{ga} \\ u\sqrt{ga} \\ v\sqrt{ga} \end{pmatrix} = \begin{pmatrix} w_1 \\ w_2 \\ w_3 \end{pmatrix}$$

At each cell interface, you have to compute the flux difference between the left (L) and right (R) states by

$$\mathbf{f}(\mathbf{v}_L) - \mathbf{f}(\mathbf{v}_R) = \sum_{p} \alpha_p c_p \mathbf{r}_p \qquad (8.52)$$

where the eigenvalues are

$$c_1 = \frac{\overline{w}_2}{\overline{w}_1} - \sqrt{\overline{w}_1^2} \qquad c_2 = \frac{\overline{w}_2}{\overline{w}_1} \qquad c_3 = \frac{\overline{w}_2}{\overline{w}_1} + \sqrt{\overline{w}_1^2} \qquad (8.53)$$

and the corresponding eigenvectors are

$$
\mathbf{r}_1 = \begin{pmatrix} \overline{w}_1 \\ \overline{w}_2 - \overline{w}_1\sqrt{\overline{w_1^2}} \\ \overline{w}_3 \end{pmatrix} \quad \mathbf{r}_2 = \begin{pmatrix} 0 \\ 0 \\ w_1 \end{pmatrix} \quad \mathbf{r}_3 = \begin{pmatrix} \overline{w}_1 \\ \overline{w}_2 + \overline{w}_1\sqrt{\overline{w_1^2}} \\ \overline{w}_3 \end{pmatrix} \tag{8.54}
$$

with coefficients

$$
\alpha_1 = \Delta w_1 - \frac{\overline{w}_1\Delta w_2 - \overline{w}_2\Delta w_1}{2\,\overline{w}_1\sqrt{\overline{w_1^2}}} \qquad \alpha_2 = \frac{\overline{w}_1\Delta w_3 - \overline{w}_3\Delta w_1}{\overline{w}_1}
$$

$$
\alpha_3 = \Delta w_1 + \frac{\overline{w}_1\Delta w_2 - \overline{w}_2\Delta w_1}{2\,\overline{w}_1\sqrt{\overline{w_1^2}}} \tag{8.55}
$$

For each variable, the notation is

$$
\Delta w = w_L - w_R, \quad \overline{w} = \tfrac{1}{2}(w_L + w_R)
$$

(note also the averaging of w_1^2). Similarly, in y-direction you get the difference between North (N) and South (S) states:

$$
\mathbf{g}(\mathbf{v}_N) - \mathbf{g}(\mathbf{v}_S) = \sum_P \beta_p c_p \mathbf{s}_p \tag{8.56}
$$

$$
\Delta w = w_N - w_S, \quad \overline{w} = \tfrac{1}{2}(w_N + w_S)
$$

$$
c_1 = \frac{\overline{w}_3}{\overline{w}_1} - \sqrt{\overline{w_1^2}} \qquad c_2 = \frac{\overline{w}_3}{\overline{w}_1} \qquad c_3 = \frac{\overline{w}_3}{\overline{w}_1} + \sqrt{\overline{w_1^2}} \tag{8.57}
$$

$$
\mathbf{s}_1 = \begin{pmatrix} \overline{w}_1 \\ \overline{w}_2 \\ \overline{w}_3 - \overline{w}_1\sqrt{\overline{w_1^2}} \end{pmatrix} \quad \mathbf{s}_2 = \begin{pmatrix} 0 \\ w_1 \\ 0 \end{pmatrix} \quad \mathbf{s}_3 = \begin{pmatrix} \overline{w}_1 \\ \overline{w}_2 \\ \overline{w}_3 + \overline{w}_1\sqrt{\overline{w_1^2}} \end{pmatrix} \tag{8.58}
$$

$$
\beta_1 = \Delta w_1 - \frac{\overline{w}_1\Delta w_3 - \overline{w}_3\Delta w_1}{2\,\overline{w}_1\sqrt{\overline{w_1^2}}} \qquad \beta_2 = \frac{\overline{w}_1\Delta w_2 - \overline{w}_2\Delta w_1}{\overline{w}_1}
$$

$$
\beta_3 = \Delta w_1 + \frac{\overline{w}_1\Delta w_3 - \overline{w}_3\Delta w_1}{2\,\overline{w}_1\sqrt{\overline{w_1^2}}} \tag{8.59}
$$

These expressions can be used both in first and second order schemes. It will be clear that the method is computationally quite involved. Truncation errors and stability are not easy to establish in the general case. It should be noted that the Roe method, as discussed here, allows unphysical "expansion shocks". A modification to prevent this is known (see LeVeque (1990) for further details).

8.7.6. Accuracy in a 1-d example

The numerical accuracy of Riemann solvers is not easily studied in terms of Fourier modes (as in chapter 9 for other methods). In order to get an impression, an example is given of a 1-d shallow-water system. You may expect difficulties for waves which are not Riemann invariants, i.e. flood waves and Rossby waves. Here, the Rossby-wave case is considered. Similar results can be expected for flood waves. The linearized 1-d SWE are

$$\frac{\partial u}{\partial t} + U \frac{\partial u}{\partial x} - fv = - g \frac{\partial h}{\partial x}$$

$$\frac{\partial v}{\partial t} + U \frac{\partial v}{\partial x} - fu = 0 \qquad (8.60)$$

$$\frac{\partial h}{\partial t} + U \frac{\partial h}{\partial x} + a \frac{\partial u}{\partial x} = \frac{\beta a}{f} v$$

with an advection velocity U and f constant. The right-hand side of (8.60c) is an artificial term leading to Rossby-type waves. If you assume a travelling wave as in section 4.4, the dispersion relation reads

$$k^3 c^3 - f^2 kc - k^3 ga\, c - \beta k\, ga = 0$$

which agrees with (4.29). If β is small, you find Poincaré waves as in (4.21); if on the other hand kc is small, you get

$$c = - \frac{\beta ga}{ga\, k^2 + f^2}$$

which is the 1-d equivalent of the Rossby wave (4.30). The amplitude- or eigenvector for this wave is approximately

$$\mathbf{v}_1 = (1, -i \frac{f}{kc}, \frac{k^2 c^2 - f^2}{k^2 gc})$$

A particular wave can be selected in the numerical simulation by specifying an initial condition with amplitude vector corresponding to just one wave type. Nonlinear terms are not included in (8.60) as they are not essential for the effect of non-derivative terms. The advantage is that you can construct exact solutions for comparison with the numerical ones. Boundary conditions are assumed periodic in x-direction, such that the wave length coincides with the length of the region.

The 1-d Riemann solver described in section 8.7.1. has been applied. The non-derivative terms have been treated by explicit forward Euler integration, i.e. they have been evaluated at the central grid point and "old" time level n. The following data were used: length $L =$ 200 km, depth $a = 20$ m, Coriolis parameter $f = 10^{-4}$ s^{-1}.

Fig. 8.6. gives the numerical and exact results after one wave period (approximately 10^4 s) for a Poincaré wave with $U = 1$ m/s, $\beta = 0$, $\Delta x = 10$ km and $\Delta t = 630$ s, resulting in a Courant number about 0.89. A very good accuracy is obtained even at this relatively coarse grid. This is more or less the ideal case for the Riemann solver as the gravity wave is one of the Riemann invariants in (8.35). The first-order upwind method (8.37) has its greatest accuracy for Courant number 1.

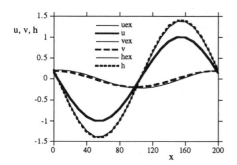

Fig. 8.6. Numerical (bold lines) and exact (thin lines) solutions for gravity (Poincaré) wave after one wave period, using first-order Riemann solver

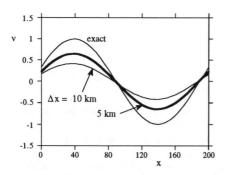

Fig. 8.7. Numerical (bold lines) and exact (thin lines) solutions for vorticity wave after one wave period, using first-order Riemann solver, at $\Delta x = 10$ km and 5 km

A vorticity wave is another Riemann invariant, but it can be run only at much smaller Courant number, the time step being restricted by the fastest (gravity) wave. Fig. 8.7 shows some results for a vorticity wave with the same data as before. The wave period is now $2 \cdot 10^5$ s, the wave speed $U = 1$ m/s and the Courant number therefore 0.063. You see that the accuracy is much worse than for the gravity wave, even at doubled resolution.

However, the corresponding results in fig. 8.8 for a Rossby wave, which is not a Riemann invariant, are very poor. Here $U = 0$, $\beta = 10^{-11}$ m^{-1}s^{-1} and the same grid size and time step have been used. Note that again a full wave period of approximately $2 \cdot 10^7$ s has been covered, which is much larger than for the previous case. Using grid sizes of 5 or 2.5 km with correspondingly reduced time steps (necessary for stability) does give some improvement but the numerical accuracy remains poor.

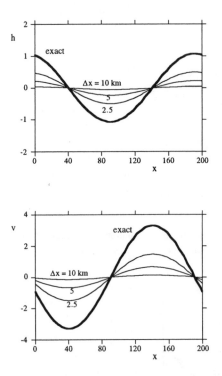

Fig. 8.8. Numerical (bold lines) and exact (thin lines) solutions for Rossby wave after one wave period, using first-order Riemann solver, at $\Delta x = 10$, 5 or 2.5 km.
Top: water level; bottom: velocity v. Units on vertical axis are arbitrary.

In order to show that it is the first-order character of the method which produces the inaccuracy, a second-order method was also used, in which the first-order method was used as a predictor, which was then corrected by a Crank-Nicolson approximation (8.10), with $A\,v^{n+1}$ taken from the predictor. This is actually a kind of second-order Runge Kutta method. This does not do justice to the idea of Riemann solvers, as the predictor-corrector method would work poorly for strongly nonlinear problems (involving shock waves), but it serves well to show the effect of the order of approximation. Fig. 8.9. shows the result for a Rossby wave (same data as in fig. 8.8). Even at coarse resolution, a very satisfactory accuracy is obtained.

As a conclusion, you see that the first-order Riemann-solver method is not sufficiently accurate for problems which are not fully described by Riemann invariants. For useful SWE results, a Riemann solver should be at least second-order accurate in regions where the solution is smooth.

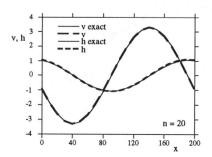

Fig. 8.9. Numerical (bold lines) and exact (thin lines) solutions for Rossby wave after one wave period, using second-order predictor-corrector method with $\Delta x = 10$ km

8.7.7. Accuracy in a 2-d example

The conclusions of the somewhat artificial 1-d example of the previous section carry over to the 2-d case. To show this, similar tests have been performed for a 2-d channel of width B and periodic in x-direction. Exact solutions are difficult to obtain for the Rossby-wave case, as this involves now a y-dependent Coriolis parameter f such that $\beta = \partial f/\partial y$. However, the solution given in section 4.4.4 can be used as a reasonable approximation.

The numerical method was the first-order component-wise Riemann solver (8.50). Data were $B = L = 800$ km, $f = 10^{-4}$, $\beta = 1.25 \; 10^{-11}$ m^{-1}s^{-1}, $a = 1$ m (to be understood as an equivalent depth for an internal wave mode), $\Delta x = 24$ km, $\Delta t = 3000$ s. The Courant number for gravity waves is approximately 0.4, which is close to its maximum in 2-d. The results for gravity and vorticity waves are comparable to those in section 8.7.6 (not shown). For the Rossby-wave case, fig. 8.10 gives cross-sections at $x = 72$ km and at $y = 72$ km at time $4.5 \; 10^6$ s. Note that, although the shape of the wave is reasonably represented, the amplitude is completely wrong. The wave has almost disappeared by numerical damping. Fig. 8.11 gives the same case with straightforward central space differences (second-order spatial accuracy) and 4-th order Runge-Kutta time integration (the latter is actually more accurate than needed and chosen just for convenience). A very satisfactory agreement is now obtained. The conclusions from the 1-d case therefore are confirmed.

It should be stressed that these tests involve smooth solutions only. Actually, most of the Riemann-solver methods have been developed for aerodynamic problems involving shock waves. The behaviour in terms of accuracy is quite different then. Such problems do occur in shallow-water flow (dam-break waves) and for such cases Riemann solvers may be a good choice. This section shows, however, that for smooth flows there is no great advantage in this type of methods.

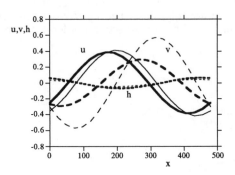

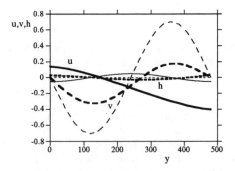

Fig. 8.10. Numerical (bold lines) and analytic (thin lines) solutions for Rossby wave in 2-d channel. Top: cross-section at $y = 72$ km, bottom: cross-section at $x = 72$ km. Values of numerical solutions have been multiplied by 1000.

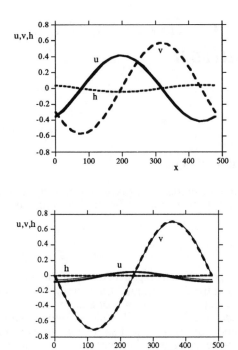

Fig. 8.11. Higher-order numerical solution for Rossby waves.
Same cross-sections as in fig. 8.10.
Bold lines numerical, thin lines analytic solutions. No scale factors applied.

Chapter 9

Effects of time discretization on wave propagation

9.1. Methods for systems of ODE's

For the standard methods of time discretization, discussed in sections 8.2 and 8.3, the numerical accuracy for linear wave propagation can be studied using the finite-difference analogue of (7.1):

$$\mathbf{v}^n = \rho^n \, exp(ik_1x + ik_2y) \, \mathbf{v}_1 \qquad (9.1)$$

(on the right-hand side, n is an exponent). The eigenvector $\mathbf{v}_1$ of the matrix $M^{-1}A$ in (8.1) is the same as in the semi-discretized continuous-time case. The error is therefore characterized by comparing ρ^n to $exp(-i\nu t)$ with $t = n \, \Delta t$ and ν is the wave frequency for the semi-discrete equations, as derived in chapter 7. Unless bottom friction plays a role (which is particularly so for flood waves), ν is real so there is no wave damping in the semi-discrete case. If there is any wave damping, it is generated by time-differencing and you can define a damping factor as

$$d_n = |\rho|^n \qquad (9.2)$$

The number of time steps is arbitrary in principle, but for purposes of comparison, it can be chosen to correspond to a physically meaningful period of time; here we choose one wave period of the semi-discrete solution:

$$n = \frac{2\pi}{\nu \Delta t}$$

The relative wave speed is defined as the ratio of the phase shifts during one time step:

$$c_r = \frac{arg(\rho)}{-\nu \, \Delta t} \qquad (9.3)$$

With the expressions for ρ in (8.3),(8.9),(8.11) (forward Euler is not included as it is unstable for this case) and taking into account that the eigenvalues of $M^{-1}A$ are $\mu = i\nu$, the quantities d_n and c_r can be evaluated. Both should be as close to unity as possible for accuracy; the deviation from unity is plotted in fig. 9.1. You can use this for gravity, vorticity and Rossby waves by inserting the appropriate value for $\nu\Delta t$ from chapter 7. Note that these errors come on top of those due to spatial discretization discussed there.

The quantity $v \Delta t$ can roughly be interpreted as $\Delta t/T$ with T the appropriate wave period. This follows from (7.6), (7.15), (7.28) for small grid sizes. The figures therefore indicate how large the time step can be taken in relation to the wave period in order to obtain a certain accuracy. As a rule of thumb, it is clear that you should have at least 10 time steps per wave period, but this depends of course on the accuracy you want to have.

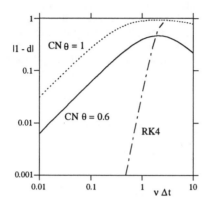

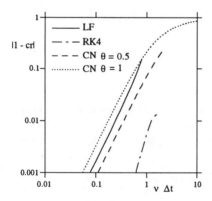

Fig. 9.1. Errors in numerical-damping factor (top) and relative wave speed (bottom) due to time-differencing for propagating waves (no physical damping)

For the leap-frog method (LF) and the Crank-Nicolson method (CN) with $\theta = 1/2$, the amplification factor has unit magnitude, so there is no numerical wave damping. However, for $\theta > 1/2$, CN gets very inaccurate in terms of wave damping. The method is then only first order in time as can be seen from the slope of the line: the error decreases linearly with Δt. For the Runge Kutta method (RK4), the error is of 4th order as expected, but nevertheless the error can become quite significant for larger values of Δt. The numerical wave speed for CN depends much less on θ than the wave damping and is second order accurate, comparable with LF. Although CN can be used for higher $v\Delta t$, being unconditionally stable, this does introduce large phase errors and should be used with caution.

In actual practice, you will have all wave types together in one model. For explicit methods, you have to satisfy the CFL condition for the largest wave speed which is the one for gravity waves. This means that $v\Delta t \approx 1$ for the fast waves and $\ll 1$ for the slow ones. The latter will then automatically be reproduced with small time-discretization errors as you can see in fig. 9.1. For implicit methods, if you are interested in the slow waves, you could violate the CFL criterion for gravity waves without getting stability problems and choose $v\Delta t$ such that you have acceptable accuracy for the slow waves. You should realize that the fast ones will then be inaccurate, which may or may not be acceptable depending on the application. You will have to make the trade-off for each particular case.

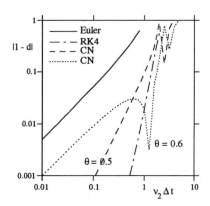

Fig. 9.2. Error in numerical damping due to time-differencing for flood waves

Only in the case of flood waves is $v = -i\, v_2$ imaginary and the damping ratio is then

$$d_n = \frac{|\rho|^n}{exp(-\,v_2\,n\,\Delta t)} = \{\,|\rho|\,exp(v_2\,\Delta t)\,\}^n \qquad (9.4)$$

No phase errors occur, which means that flood waves will move with the correct speed. The damping factor is illustrated in fig. 9.2 for various methods. The number of time steps is chosen to correspond to one relaxation time. In this case $v_2\Delta t$ is close to $\Delta t/T_r$ (7.14). The time step apparently should be small relative to the relaxation time. Forward Euler can be used but is very inaccurate and therefore not advisable. LF is unstable in this case. CN is quite accurate for $\theta = 1/2$ again, but otherwise serious numerical damping is introduced. RK4 is 4th order accurate but the error can get large for large time steps.

9.2. Gravity waves

In the methods of sections 8.4 to 8.6, time discretization is mixed with spatial discretization, e.g. by treating some terms explicitly and others implicitly. The time-discretization error can then not be determined separately as in section 9.1. Fortunately, it turns out that it is often not very much influenced by spatial discretization, so by fixing the latter to a reasonable resolution, you can get a fairly good impression of the time discretization error. This may work out differently for the various types of waves; in this section gravity waves are discussed ($r = 0$, $U = V = 0$).

To make a comparison, take the same grid for all methods, say the C-grid which is most commonly used. The results will not be essentially different for the other grids. Let us fix the resolution at $\xi = k_1 \Delta x = \eta = k_2 \Delta y = \pi/10$ (that is: 20 grid points per wave length in each direction). The semi-discrete solution is then (7.1) with

$$v \, \Delta t = k \sqrt{ga} \, \Delta t \left(-\frac{D_{0x}^2 + D_{0y}^2}{k^2} + \frac{(M_x M_y)^2}{(Rk)^2} \right)^{1/2} = \tau \left(\frac{sin^2 \xi}{\xi^2} + \frac{cos^4 \xi}{(Rk)^2} \right)^{1/2} \qquad (9.5)$$

where $\tau = \sqrt{ga} \, k \, \Delta t \sim \Delta t/T$. Let us take $Rk = 1$ in the following examples; this means that the wave length is 2π times the Rossby radius.

Specializing (8.12) to a C-grid for gravity waves gives

$$\frac{u^{n+1} - u^n}{\Delta t} - f M_x M_y v^n + g D_{0x} h^n = 0$$

$$\frac{v^{n+1} - v^n}{\Delta t} + f M_x M_y u^{n+1} + g D_{0y} h^n = 0 \qquad (9.6)$$

$$\frac{h^{n+1} - h^n}{\Delta t} + a (D_{0x} u^{n+1} + D_{0y} v^{n+1}) = 0$$

Substituting a solution of the form (9.1), you find that the following determinant should vanish in order to have a non-trivial solution (note the ρ's corresponding to terms evaluated at time level $n+1$)

$$\begin{vmatrix} \rho - 1 & -f \Delta t \, M_x M_y & g \Delta t \, D_{0x} \\ \rho f \Delta t \, M_x M_y & \rho - 1 & g \Delta t \, D_{0y} \\ \rho a \Delta t \, D_{0x} & \rho a \Delta t \, D_{0y} & \rho - 1 \end{vmatrix} = 0$$

or

$$(\rho - 1)(\rho^2 - 2\gamma\rho + 1) = 0 \qquad (9.7)$$

with $\gamma = 1 + \frac{1}{2} ga \, \Delta t^2 (D_{0x}^2 + D_{0y}^2) - (f \Delta t \, M_x M_y)^2 - f \Delta t \, M_x M_y ga \, \Delta t^2 D_{0x} D_{0y}$

Eq.(9.7) has one solution $\rho = 1$ corresponding to vorticity transport (which would be stationary without advection, but see section 9.3). The other two solutions correspond to gravity waves propagating in opposite directions:

$$\rho = \gamma \pm i \sqrt{1 - \gamma^2} \qquad\qquad (9.8)$$

For the chosen conditions,

$$\gamma = 1 - \frac{1}{2}\tau^2 \frac{sin^2\xi + sin^2\eta}{(k\,\Delta x)^2} - \left(\frac{\tau\cos\xi\cos\eta}{Rk}\right)^2 + \frac{\tau^3\cos\xi\cos\eta}{Rk}\frac{sin\,\xi\,sin\,\eta}{(k\,\Delta x)^2} =$$

$$= 1 - \frac{1}{2}\tau^2\frac{sin^2\xi}{\xi^2} - \left(\frac{\tau\cos^2\xi}{Rk}\right)^2 + \frac{\tau^3}{Rk}\frac{sin^2\xi\cos^2\xi}{2\xi^2}$$

$$\qquad\qquad (9.9)$$

for $\xi = \eta$ and $k^2\Delta x^2 = k_1^2\Delta x^2 + k_2^2\Delta x^2 = \xi^2 + \eta^2 = 2\xi^2$.

From (9.8) you can compute the same numerical quantities (9.2) and (9.3). These are illustrated in fig. 9.3 together with those for other methods. There is no wave damping in this case if $|\gamma| \le 1$. The wave speed shows a first-order error. You can check this by observing that for $f = 0$ and $\xi \to 0$, (9.8) gives $\rho \approx 1 \pm i\tau$ which is only a first-order approximation of the required value $exp(\pm i\tau)$. Even though the errors are not very large, the conclusion is that there is a sort of mismatch with the (second-order) spatial discretization error shown in fig. 7.1 for grid C.

The semi-implicit leap-frog method (8.19) differs from the standard LF method only in the frictional terms, which are disregarded here, so this method is not separately shown. For gravity waves, both methods are therefore equivalent (see fig. 9.1). The second semi-implicit leap-frog method (8.21) gives a determinant equation

$$\begin{vmatrix} \rho^2 - 1 & -2\rho\,f\Delta t\,M_xM_y & g\Delta t\,D_{0x}(\rho^2+1) \\ 2\rho f\Delta t\,M_xM_y & \rho^2 - 1 & g\Delta t\,D_{0y}(\rho^2+1) \\ a\Delta t\,D_{0x}(\rho^2+1) & a\Delta t\,D_{0y}(\rho^2+1) & \rho^2 - 1 \end{vmatrix} = 0$$

or

$$(\rho^2 - 1)(\rho^4 - 2\gamma\rho^2 + 1) = 0 \qquad\qquad (9.10)$$

with now a different coefficient γ

$$\gamma = \frac{1 + ga\,\Delta t^2(D_{0x}^2 + D_{0y}^2) - 2(f\,\Delta t\,M_xM_y)^2}{1 - ga\,\Delta t^2(D_{0x}^2 + D_{0y}^2)} = \frac{1 - \tau^2\frac{sin^2\xi}{\xi^2} - 2\left(\frac{\tau\cos^2\xi}{Rk}\right)^2}{1 + \tau^2\frac{sin^2\xi}{\xi^2}}$$

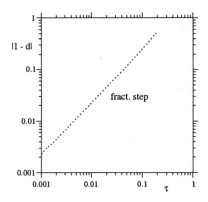

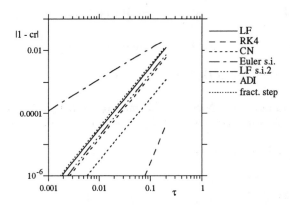

Fig. 9.3. Errors in damping (top) and wave speed (bottom) due to combined time- and spatial differencing for gravity waves. Methods other than fractional step do not have amplitude errors

There are 6 roots to (9.10), 3 of which correspond to physical waves and the remaining ones are spurious waves, as usual with LF type methods. The root $\rho = 1$ again refers to the vorticity wave. For stability, $|\gamma| \leq 1$ is required, similar to (8.23); then there is no wave damping. The wave speed shows a second-order accuracy.

For the ADI method (8.25), the analysis is technically somewhat more complicated. The idea is to eliminate the auxiliary variables $\mathbf{v}^*$ and then substitute the tentative solution (9.1). The result is again a root $\rho = 1$ (vorticity transport) and two roots corresponding to the gravity waves:

$$\rho = \frac{1 \pm i\,\gamma/2}{1 \pm (-i\,\gamma/2)} \tag{9.11}$$

where γ has again a different meaning

$$\gamma^2 = \frac{\tau^2 \dfrac{\sin^2 \xi}{\xi^2} + \left(\dfrac{\tau \cos^2 \xi}{Rk}\right)^2 + \tau^4 \dfrac{\sin^4 \xi}{16\xi^4}}{1 - \dfrac{1}{4}\left(\dfrac{\tau \cos^2 \xi}{Rk}\right)^2}$$

There is again no wave damping. Fig. 9.3 shows that the wave speed has second-order accuracy and is better than for leap-frog schemes. It is even better than the standard Crank-Nicolson scheme from which it is derived. Apparently, the additional terms generated in the splitting have a favorable effect on the phase error.

Finally, the fractional-step method (8.32) is not so easily translated to a C-grid, so we give the result for an E-grid. The determinant equation becomes

$$\begin{vmatrix} \rho - 1 & -f\Delta t & \frac{1}{2}g\Delta t\, D_{0x}(\rho+1) \\ f\Delta t & \rho - 1 & \frac{1}{2}g\Delta t\, D_{0y}(\rho+1) \\ \frac{1}{2}a\Delta t\, D_{0x}(\rho+1) & \frac{1}{2}a\Delta t\, D_{0y}(\rho+1) & \rho - 1 \end{vmatrix} = 0$$

This corresponds to the Crank-Nicolson scheme except for the Coriolis terms which are treated explicitly. There is again a root $\rho = 1$; the other roots follow from

$$(1+p)\,\rho^2 - 2(1-p)\,\rho + 1 + p + q = 0 \tag{9.12}$$

with

$$p = \tau^2 \frac{\sin^2 \xi}{4\,\xi^2} \qquad q = \left(\frac{\tau}{Rk}\right)^2$$

The roots should be compared with $v\,\Delta t$ from section 7.2. As an exception, the roots do not have unit magnitude, so there is a numerical wave damping as shown in fig. 9.3a. The error is of first order in the time step and can be relatively serious for larger time steps. This is completely explained by the explicit treatment of the Coriolis terms. If $f = 0$, the amplitude error disappears. The wave speed is second-order accurate in any case.

In fig. 9.4 the case without Coriolis terms ($f = 0$) is shown. All methods (except RK4) then have an amplification factor with unit amplitude and consequently no numerical wave damping. The first-order error in the semi-implicit Euler method has disappeared and actually all methods are relatively close together. The ADI and the second semi-implicit LF methods are are somewhat less accurate than in fig. 9.3; the converse is true for the fractional-step method.

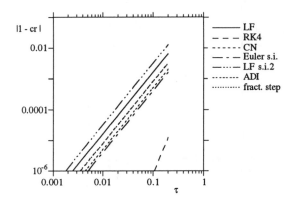

Fig. 9.4. Errors in wave speed due to combined time-and spatial differencing for gravity waves without Coriolis terms

9.3. Vorticity waves

For the propagation of vorticity waves which were represented by a root $\rho = 1$ in the previous section, you have to reintroduce the advection velocity and look at the treatment of the advective terms in finite-difference form. For example, the standard leap-frog method, (8.2) would give, disregarding water-level variations and assuming advection only in x-direction for simplicity

$$u^{n+1} - u^{n-1} + 2 \, \Delta t \, U \, M_x D_{0x} u^n = 0 \qquad (9.13)$$

where the central space-difference is effectively taken over a double interval on the staggered C-grid. A solution of the form (9.1) is possible if

$$\rho = \sqrt{1 - \gamma^2} - i \, \gamma \qquad (9.14)$$

with

$$\gamma = U \frac{\Delta t}{2\Delta x} \sin 2\xi = \tau \frac{\sin 2\xi}{2\xi}$$

and $\tau = U k_1 \Delta t$ indicates the ratio of the time step to the period of the vorticity wave. The corresponding phase error by comparison with the semi-discrete factor $exp(-iv \, \Delta t)$ from (7.17) is illustrated in fig. 9.5. together with results for the other methods. Similar results are obtained for the RK4 and CN methods. The forward Euler method is unstable for advection only and is therefore not included.

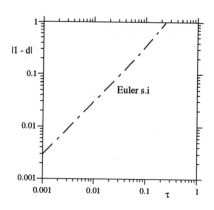

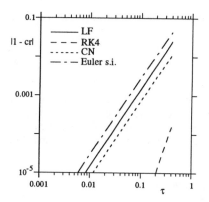

Fig. 9.5. Errors in damping (top) and wave speed (bottom) due to combined time-and spatial differencing for vorticity waves

For the Euler semi-implicit method on a C-grid, you would get from (8.12)

$$\rho = 1 - i\tau \frac{\sin 2\xi}{2\xi} \tag{9.15}$$

This gives not only a phase error, but an amplitude error as well (fig. 9.5a) which can be quite serious. Interestingly, the phase error is still second-order accurate. Both semi-implicit leap-frog methods treat advection exactly the same way as standard leap-frog, so separate results need not be given.

For the ADI method, (8.25) gives with elimination of the intermediate step:

$$(1 + \tfrac{1}{2} \Delta t \, U \, M_x D_{0x}) u^{n+1} = (1 - \tfrac{1}{2} \Delta t \, U \, M_x D_{0x}) u^n \tag{9.16}$$

which is the same as Crank-Nicolson. Finally, the fractional step method (8.32) produces exact results for advection in case of a constant advection speed (assuming negligible errors from high-order spatial interpolation).

Fig. 9.5 shows (except for the semi-implicit Euler method) that advection of vorticity is represented at about the same accuracy as gravity-wave propagation (compare fig. 9.3, 9.4). The Runge-Kutta method again performs significantly better than any of the other methods unless the time step is large. However, you should keep in mind that a spatial discretization error remains (ch. 7).

9.4. Flood waves

For typical flood waves, you can neglect Coriolis and inertia terms in the momentum equations, but retain bottom friction as a dominant term (section 4.4.3). As in (9.4), flood wave damping due to time discretization is characterized by comparison with its semi-discrete counterpart (7.23). There is no difference between A,C and E grids.

Consider again the semi-implicit Euler method, which becomes with the current assumptions

$$g \, D_{0x} h^n + r \, u^n = 0$$
$$g \, D_{0y} h^n + r \, v^n = 0 \tag{9.17}$$
$$h^{n+1} - h^n + a \Delta t \, \{D_{0x} u^{n+1} + D_{0y} v^{n+1}\} = 0$$

This gives the determinant equation

$$\begin{vmatrix} r & 0 & g D_{0x} \\ 0 & r & g D_{0y} \\ \rho a \Delta t \, D_{0x} & \rho a \Delta t \, D_{0y} & \rho - 1 \end{vmatrix} = 0$$

or

$$\rho = \left\{ 1 - \frac{ga \, \Delta t}{r} (D_{0x}^2 + D_{0y}^2) \right\}^{-1} = \left\{ 1 + \tau \frac{\sin^2 \xi + \sin^2 \eta}{\xi^2 + \eta^2} \right\}^{-1} \tag{9.18}$$

where $\tau = ga \, k^2 \Delta t / r$ is the ratio between time step and damping time scale.

The standard leap-frog method would be unstable for this case. The first semi-implicit LF method (8.19) has been stabilized by treating the frictional terms implicitly. This gives

$$\begin{vmatrix} \frac{1}{2}r\,(\rho^2+1) & 0 & \rho g D_{0x} \\ 0 & \frac{1}{2}r\,(\rho^2+1) & \rho g D_{0y} \\ 2\rho a \Delta t\, D_{0x} & 2\rho a \Delta t\, D_{0y} & \rho^2 - 1 \end{vmatrix} = 0$$

or

$$\rho^4 - 4\tau\rho^2\,(D_{0x}^2 + D_{0y}^2) - 1 = 0$$

so

$$\rho = \left(\sqrt{1 + 4\,\gamma^2} - 2\gamma\right)^{1/2} \tag{9.19}$$

with

$$\gamma = \tau\,\frac{sin^2\xi + sin^2\eta}{\xi^2 + \eta^2}$$

For the second semi-implicit LF method (8.21), you obtain

$$\begin{vmatrix} \rho r & 0 & \frac{1}{2}g D_{0x}(\rho^2+1) \\ 0 & \rho r & \frac{1}{2}g D_{0y}(\rho^2+1) \\ a\Delta t\, D_{0x}(\rho^2+1) & a\Delta t\, D_{0y}(\rho^2+1) & \rho^2 - 1 \end{vmatrix} = 0$$

or

$$\gamma\rho^4 + 2\,\rho^3 + 2\gamma\rho^2 - 2\rho + \gamma = 0 \tag{9.20}$$

with the same γ as in (9.19). It is not easy to solve this equation. At any rate, it is clear that it has a number of spurious roots. You can, however, approximate the physically relevant root for small τ (or small γ) by assuming $\rho = 1 + \rho_1\gamma + \rho_2\gamma^2$, substituting and equating equal powers of γ, you find then $\rho \approx 1 - \gamma + \gamma^2/3$.

For the ADI method, eliminating the intermediate step from (8.25) and neglecting inertia terms gives for the present case

$$\begin{vmatrix} \frac{1}{2}r(\rho+1) & \frac{1}{4}ga\,\Delta t\, D_{0x}D_{0y}(\rho-1) & \frac{1}{2}g D_{0x}(\rho+1) \\ 0 & \frac{1}{2}r(\rho+1) & \frac{1}{2}g D_{0y}(\rho+1) \\ \frac{1}{2}a\Delta t\, D_{0x}(\rho+1) & \frac{1}{2}a\Delta t\, D_{0y}(\rho+1) & \rho - 1 \end{vmatrix} = 0$$

Dividing each equation by $\rho+1$ gives

$$\frac{\rho - 1}{\rho + 1} = \frac{\dfrac{ga\,\Delta t}{2r}(D_{0x}^2 + D_{0y}^2)}{1 + \left(\dfrac{ga\,\Delta t}{2r}D_{0x}D_{0y}\right)^2} = \frac{-\dfrac{1}{2}\tau\,\dfrac{sin^2\xi + sin^2\eta}{\xi^2 + \eta^2}}{1 + \left(\dfrac{\tau}{2}\dfrac{sin\,\xi\,sin\,\eta}{\xi^2 + \eta^2}\right)^2} \tag{9.21}$$

Finally, for the fractional-step method (8.32), you will find exactly the same as for the standard Crank-Nicolson method.

The relative error in the damping factor is shown in fig. 9.6 for $\xi = \eta = \pi/10$. The standard forward Euler method can be used here (due to the presence of friction) and it is found to have even second-order accuracy, together with its semi-implicit companion and the second semi-implicit LF method. At small time steps, a group of methods (first semi-implicit LF, ADI/CN) produce more accurate results, even to third order. The RK4 method, as it should, gives 4th order and very small errors. Again, you can observe that at coarse time steps, higher-order methods do not necessarily perform better than lower-order ones.

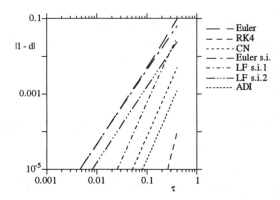

Fig. 9.6. Errors in wave damping due to combined time- and spatial differencing for flood waves

9.5. Rossby waves

The quasi-geostrophic model equations for Rossby waves are the potential-vorticity equation, derived with *variable f* parameter, augmented by the momentum equations in which the f parameter is then taken constant (see section 4.4.4 eqs.(4.32)). To see how numerical approximations work, you will have to construct the numerical equivalent of the potential vorticity equation (4.32c). As a first example, consider the semi-implicit Euler method. On the C-grid, the momentum equations are given by (9.6). Taking the discrete curl of these gives

$$(D_{0x}v - D_{0y}u)^{n+1} - (D_{0x}v - D_{0y}u)^n + f\Delta t\, M_xM_y(D_{0x}u^{n+1} + D_{0y}v^n) + \beta\Delta t\, M_xM_y^2 v^n = 0$$

Note that the divergence term is not defined at one single time level; eliminating the divergence at level $n+1$ using (9.6c) gives

$$(D_{0x}v - D_{0y}u)^{n+1} - (D_{0x}v - D_{0y}u)^n - \frac{fM_xM_y}{a}(h^{n+1} - h^n) +$$
$$- f\Delta t\, M_xM_y\, D_{0y}(v^{n+1} - v^n) + \beta\Delta t\, M_xM_y^2 v^n = 0 \qquad (9.22)$$

(compare with (7.29)). Now, you can take f and β constant in (9.6 a,b and 9.22). Wave propagation is then governed by

$$\begin{vmatrix} 0 & -f\,M_xM_y & g\,D_{0x} \\[2mm] \rho f\,M_xM_y & 0 & g\,D_{0y} \\[2mm] -(\rho-1)D_{0y} & a_{32} & -\dfrac{fM_xM_y}{a}(\rho-1) \end{vmatrix} = 0$$

where $a_{32} = (\rho-1)D_{0x} + \beta\Delta t M_xM_y^2 - f\Delta t M_xM_yD_{0y}(\rho-1)$.
Elaborating the determinant yields a complicated quadratic equation for ρ:

$$\rho(\rho - 1)(a - \tau d) + (\rho - 1)b + \tau\rho c = 0 \qquad (9.23)$$

where

$$a = D_{0x}^2 - \frac{(fM_xM_y)^2}{ga} \qquad b = D_{0y}^2 \qquad c = \frac{k^2}{k_1}D_{0x}M_xM_y^2 \qquad d = \frac{fk^2}{\beta k_1}D_{0y}M_xM_y$$

and $\tau = \beta\Delta t k_1/k^2$ indicates the ratio of time step to wave period. You should note that a term proportional to $f\Delta t$ occurs in the coefficient d, which should not be there (compare

with the continuous and semi-discrete counterparts 4.30 and 7.30). Even though it is proportional to Δt, this term need not be small because

$$f\Delta t = \frac{fk^2}{\beta k_1}\tau \sim \frac{2\pi f r_0}{fL}\tau$$

(L = wave length, r_0 = radius of the earth) and the coefficient in front of τ may be quite large. Fig. 9.7 gives the physically relevant root of (9.23) for various values of $f\Delta t/\tau$. Significant errors are seen in the wave speed which drops from second to first order if $f\Delta t \neq 0$. It turns out that wave damping is hardly affected (not shown). For the comparison with other methods see fig. 9.8 where $f\Delta t = 0$. Even then, you see that a relatively large, first order error in wave damping occurs. The accuracy of the wave speed is much better and of second order.

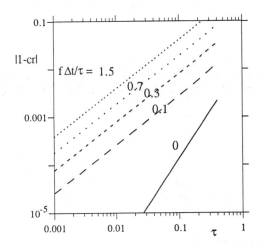

Fig. 9.7. Errors in wave speed for Rossby waves in the Euler semi-implicit method due to combined time- and spatial differencing

The first semi-implicit leap-frog method is identical to standard leap-frog in this case. The second semi-implicit leap-frog method gives as a vorticity equation

$$(D_{0x}v - D_{0y}u)^{n+1} - (D_{0x}v - D_{0y}u)^{n-1} + 2f\Delta t\, M_xM_y(D_{0x}u^n + D_{0y}v^n) +$$
$$+ 2\beta\Delta t\, M_xM_y^2v^n = 0$$

To eliminate the divergence, you should average this equation for values of n replaced by $n+1$ and $n-1$ and then apply (8.21c):

$$(D_{0x}v - D_{0y}u)^{n+2} - (D_{0x}v - D_{0y}u)^{n-2} - \frac{2fM_xM_y}{a}(h^{n+1} - h^{n-1}) +$$
$$+ 2\,\beta\Delta t\, M_xM_y^2(v^{n+1}+v^{n-1}) = 0 \qquad (9.24)$$

Then, with (8.21a,b), the determinant equation becomes

$$\begin{vmatrix} 0 & -\rho f\, M_xM_y & \frac{1}{2}(\rho^2+1)\, g\, D_{0x} \\ \rho f\, M_xM_y & 0 & \frac{1}{2}(\rho^2+1)\, g\, D_{0y} \\ -(\rho^4-1)D_{0y} & (\rho^4-1)D_{0x}+ 2\beta\Delta t\,(\rho^2+1)M_xM_y^2 & -2fM_xM_y/a\,(\rho^2-1) \end{vmatrix} = 0$$

or

$$(\rho^2+1)(\rho^4-1)\frac{D_{0x}^2+D_{0x}^2}{4k^2} - \frac{f^2 M_x^2 M_y^2}{ga\,k^2}\rho(\rho^2-1) + \frac{1}{2}\tau(\rho^2+1)\frac{2D_{0x}}{k_1}M_x M_y^2 = 0$$

$$(9.25)$$

Obviously, this equation has a number of spurious roots. The physically relevant root has been solved up to second order in τ (details are not shown). Fig. 9.8 indicates that the method produces an even higher wave damping than the Euler semi-implicit method. The wave-speed errors are also significantly larger. Apparently, the way of handling the Coriolis terms in this method is not very fortunate for Rossby waves which depend essentially on these terms.

The derivation for the ADI method is technically involved but straightforward. Intermediate values are eliminated from (8.25) and then the discrete vorticity equation is formed. Due to the ADI splitting, a number of terms of the form $\Delta t^2(v^{n+1}-v^n)$ show up. These are of order Δt^3 and can therefore be assumed not to influence the wave propagation as a first approximation. Disregarding these terms gives

$$\rho = \frac{1+\gamma}{1-\gamma} \qquad \gamma = \frac{\frac{1}{2}\beta\Delta t\,M_x M_y^2 D_{0x}}{-(D_{0x}^2+D_{0y}^2)+f^2 M_x^2 M_y^2/ga} = -\frac{1}{2}i\,v\,\Delta t \qquad (9.26)$$

The amplification factor has unit amplitude so there is no wave damping (this is because $\theta = 1/2$ has been assumed). The remaining expression (9.26) is actually identical to the one for the Crank-Nicolson method.

Finally, for the fractional step method (on the E-grid), (8.32) leads to the vorticity equation

$$(D_{0x}v - D_{0y}u)^{n+1} - (D_{0x}v - D_{0y}u)^n + f\Delta t\,(D_{0x}u^n + D_{0y}v^n) + \beta\Delta t\,M_y v^n = 0$$

Averaging this for values n and $n+1$ and substituting (8.32c) gives

$$(D_{0x}v - D_{0y}u)^{n+2} - (D_{0x}v - D_{0y}u)^n - \frac{2f}{a}(h^{n+1}-h^n) + \beta\Delta t\,M_y(v^{n+1}+v^n) = 0$$

$$(9.27)$$

The determinant equation is

$$\begin{vmatrix} 0 & -f & \frac{1}{2}(\rho+1)\,g\,D_{0x} \\ f & 0 & \frac{1}{2}(\rho+1)\,g\,D_{0y} \\ -(\rho^2-1)D_{0y} & (\rho^2-1)D_{0x}+\beta\Delta t\,(\rho+1)M_y & -\frac{2f}{a}(\rho-1) \end{vmatrix} = 0$$

or

$$a(\rho-1)(\rho+1)^2 + 4b\,(\rho-1) - \tau c(\rho+1)^2 = 0 \qquad (9.28)$$

with

$$a = -\frac{D_{0x}^2 + D_{0y}^2}{k^2} \qquad b = \frac{1}{(Rk)^2} \qquad c = \frac{D_{0x}}{k_1} M_y$$

Eq.(9.28) has been solved for small τ up to second order. Fig. 9.8 shows that there is some, but not very serious, wave damping. The wave speed is as accurate as that for the leap-frog method.

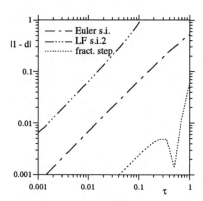

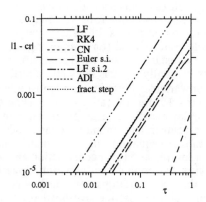

Fig. 9.8. Errors in wave damping (top) and wave speed (bottom) due to combined time- and spatial differencing for Rossby waves ($f\Delta t = 0$)

9.6. Amount of work

If you want to compare the performance of various numerical methods to solve a certain problem, one factor is the resolution required to obtain a specified numerical accuracy. This has been discussed at length in chapter 7 and the previous sections. However, there is another factor: the amount of work per grid point and time step. It is not very interesting what accuracy can be reached on a given grid and at a given time step. The important question is: what is the work needed to obtain a solution at a given accuracy?

The amount of work is defined as the number of arithmetical operations (multiplications, divisions, square roots; additions are disregarded) to advance the solution on one grid point over one time step. The number of operations is estimated for the complete nonlinear shallow-water equations, however without horizontal diffusion terms. The numbers given below do not come from computer tests but from a consideration of the finite-difference (or finite-element) expressions. In actual implementations, certain variations may occur, so the numbers given here should be considered estimates, though realistic ones. Vectorization of computer codes has not been taken into account. The type of computation involved in these methods is generally well amenable to vectorization and you can assume the level of vectorization to be roughly the same for all methods. Then the number of operations is still a good basis for comparison.

First of all, consider the leap-frog method applied to (6.9) on grid C. The equation of continuity takes 5 multiplications per time step. Each of the momentum equations requires 14 multiplications, so the total number of operations is 33. Several other methods differ from this method only in the way of time stepping, with the same work per time step: forward Euler and its semi-implicit variant, and the first semi-implicit leap-frog method.

For the 4-th order Runge-Kutta method, each of the four stages requires the same 33 operations, but the RK procedure itself (8.8) takes an additional 15 operations so the total work for one complete time step becomes about 150. You may also consider the RK4 method applied to a spatial discretization that is 4th-order accurate as well. One such method is the consistent finite-element method (on a regular grid, (6.17) and (6.20)). Here, the element matrices M are constant and have to be constructed only once. If the number of time steps is large (as it usually is), this does not contribute to the cost. The matrices D^x, D^y, however, work on variable coefficients and must therefore be built up in each time step again. This requires 15 multiplications each. Then these matrices are multiplied by the vector of unknowns; per triangular element this requires 9 multiplications. Taking everything together, the continuity equation takes 98 operations per time step per triangle. In the momentum equations, the same matrices D occur; new ones are the N matrices which take 28 multiplications for construction and 9 for multiplication by the vector of unknowns. Taking two momentum equations together, we count 202 multiplications so the total number of operations for one level of the Runge-Kutta procedure is 300. There are four such levels; the overhead for the RK procedure is negligible, so you find about 1200 operations per triangle per time step. We assume that the number of triangles is about the same as the number of grid points. If you use consistent mass matrices, 3 systems of equations of the type $Mx = b$ have to be solved. As shown below, this takes $3*(2n_x)$ operations per grid point where n_x is the linear number of grid points. The total number for consistent FEM and RK4 then becomes $1200 + 6n_x$ per grid point. This is a quite high number which explains why the finite-element method is generally considered expensive.

The remaining methods are implicit and require the solution of a system of equations per time step. In the Crank-Nicolson method, a full system of 3 times the number of grid points has to be built up and solved in each time step. This can only be reasonably done using iterative techniques but it is difficult to estimate the number of iterations needed, so we will rather consider the ADI variant. There, tridiagonal systems have to be solved for each row and column separately. Computing the right-hand sides is about the same amount of work as one explicit time step (i.e. 33 operations). The number of operations for one tridiagonal system with n_x unknowns is $5n_x$, and there are $2n_x$ of such systems, so total work is $10 n_x^2$ or 10 per grid point. In total, you find $2*33+10 \approx 75$ operations per grid point. This is a lower limit; in more elaborate versions like Stellings' (1983) the number is significantly higher.

In the second semi-implicit leap-frog method and in the fractional-step method, a Helmholtz equation for the water level of the type (8.22) must be solved. This can be arranged such that the matrix is constant by only taking the average water level into account and including the variations of water level in the right-hand side. Then, an *LU* decomposition can be made once and for all and in each time step, only multiplications by L and U are needed, each of which takes the number of grid points n_x^2 times the band width n_x. Computing the right-hand side is again equivalent to one explicit step, so the total work becomes $2n_x^3 + 33 n_x^2$ or $2n_x + 33$ per grid point. In the fractional-step method, additional work is needed for accurate solution of the frictional effect and the advection along characteristics. This is difficult to estimate but let us assume that it takes of the order of 100 additional operations per grid point.

Summarizing, you get the following estimates.

Table 9.1. Number of operations per grid point

time integration	grid	operation count w
leap-frog	C	33
forward Euler	C	33
RK4	C	150
RK4	FEM	$1200 + 6n_x$
Crank-Nicolson	C	depends on iteration
Euler semi-implicit	C	33
leap-frog semi-implicit 1	C	33
leap-frog semi-implicit 2	C	$33 + 2n_x$
ADI	C	75
fractional step	E	$130 + 2n_x$

9.7. Evaluation

There is a variety of numerical methods to solve the SWE with greatly different accuracies. In order to compare them, you should set a desired accuracy (which comes from the application; no general guidelines are possible) and then select the method producing a solution with this accuracy at a minimum cost. It is therefore necessary to specify the amount of work needed to generate a numerical solution at a certain accuracy. With some schematization, you could do this as follows.

From chapter 7 and this chapter, you can see that the numerical error in either wave damping or wave speeed can to a fairly good approximation be described as

$$\text{error} = a\ \xi^m + b\ \tau^n \le \varepsilon$$

where ε is the required accuracy (in either wave damping or speed), ξ, τ are the ratios of grid size to wave length and time step to wave period respectively. The coefficients a,b,m,n are different for each numerical method and can be read from the accuracy graphs given previously. In fig. 9.9, a drawn line represents combinations of ξ, τ producing a constant error ε. The work w per time step and grid point has been estimated in the preceding section and is given in table 9.1. The total amount of work W to cover a certain spatial area and a period of time is proportional to $(L/\Delta x)^2(T/\Delta t)$. Here, it has been assumed that $\Delta x = \Delta y$. The constant of proportionality is the same for all methods because it just says how many wave-lengths and wave periods have to be covered; therefore this constant can be omitted for a comparison. Then

$$W = \frac{w}{\xi^2 \tau}$$

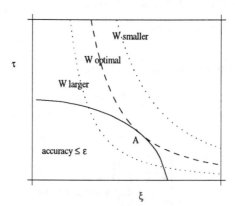

Fig. 9.9. Accuracy and work as functions of spatial and temporal resolution.
Drawn line: constant accuracy ε, dotted and dashed lines: constant total work.

Lines of constant total work are included in fig. 9.9. The combination producing the solution at the specified accuracy using the least amount of work is then given by the tangent point A in fig. 9.9 where two relations have to be satisfied :

$$\text{(intersection)} \quad \tau = \frac{w}{\xi^2 W} = \left(\frac{\varepsilon - a\xi^m}{b}\right)^{1/n} \tag{9.29}$$

(tangency) $\quad \dfrac{-2w}{\xi^3 W} = \dfrac{-ma\xi^{m-1}}{nb}\left(\dfrac{\varepsilon - a\xi^m}{b}\right)^{1/n-1}$

With some algebra, this gives

$$\xi = \left(\dfrac{n\varepsilon}{a\left(\frac{1}{2}m + n\right)}\right)^{1/m} \qquad\qquad (9.30)$$

and W/w and τ can be determined from (9.29). In this way, you can find the total work as a function of accuracy.

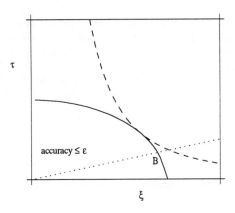

Fig. 9.10. Accuracy and work as functions of spatial and temporal resolution constrained by stability condition (dotted line)

However, two kinds of constraints may be present which prevent this optimal value to be reached. Firstly, there may be a *stability constraint* of the form $\tau/\xi \le c$ (this is discussed below), indicated by a dotted line in fig. 9.10. Then, the optimum point is B in fig. 9.10 and you find

$$c\xi = \left(\dfrac{\varepsilon - a\xi^m}{b}\right)^{1/n}$$

or

$$bc^n\xi^n + a\,\xi^m = \varepsilon \qquad\qquad (9.31)$$

You cannot easily solve this for general m and n. However, it is sufficient to have a good estimate of the solution. There are three possibilities:
(i) Most commonly, the orders of spatial and temporal approximations will be the same, so $m = n$; then

$$\xi = \left(\frac{\varepsilon}{a + b \, c^n}\right)^{1/n}$$

(ii) $m > n$; then the m-th power may be neglected for small ξ, so

$$\xi \approx \frac{1}{c}\left(\frac{\varepsilon}{b}\right)^{1/n}$$

(iii) $m < n$; then conversely

$$\xi \approx \left(\frac{\varepsilon}{a}\right)^{1/m}$$

In each case, you find τ from the stability condition $\tau = c\xi$ and W/w from its definition.

The second constraint may be the resolution. It is quite common that you will choose a finer resolution (smaller ξ) than you would strictly need just from accuracy analysis, e.g. in order to represent bottom topography or the shape of coastlines satisfactorily. For example, for tidal waves in the sea, you might find that a grid size of 100 km would be sufficient for accuracy of wave propagation, but to represent bottom and coastal topography, you might decide to use a grid size of 10 km. Then you have a constraint $\xi \leq \xi_0$ and the optimum will be at point C in fig. 9.11:

$$\tau = \left(\frac{\varepsilon - a\xi_0^m}{b}\right)^{1/n} \tag{9.32}$$

Finally, both constraints may occur simultaneously. Then the optimum is in point D in fig. 9.12 with $\xi = \xi_0$, $\tau = c\xi_0$.

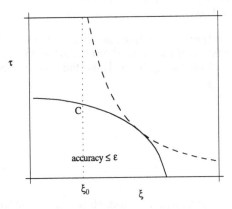

Fig. 9.11. Accuracy and work as functions of spatial and temporal resolution constrained by resolution (dotted line)

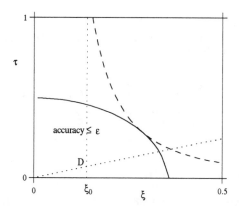

Fig. 9.12. Accuracy and work as functions of spatial and temporal resolution constrained by stability condition and resolution (dotted lines)

A stability constraint usually is in the form of a Courant number restriction:

$$\sqrt{ga}\,\frac{\Delta t}{\Delta x} \le \sigma_0$$

the value of the right-hand side depending on the type of discretization (ch. 8). For the various wave types, with different definitions of τ (see sections 9.2 to 9.5), this can be translated as follows. For gravity waves

$$\sqrt{ga}\,\frac{\Delta t}{\Delta x} = \frac{\tau}{\xi\sqrt{2}} \le \sigma_0 \tag{9.33}$$

For vorticity waves

$$\sqrt{ga}\,\frac{\Delta t}{\Delta x} = \frac{\tau}{\xi F} \le \sigma_0, \qquad F = \frac{U}{\sqrt{ga}} \text{ Froude number} \tag{9.34}$$

For flood waves

$$\sqrt{ga}\,\frac{\Delta t}{\Delta x} = \sqrt{\frac{r^2}{2gak^2}}\,\frac{\tau}{\xi} \le \sigma_0 \tag{9.35}$$

For Rossby waves

$$\sqrt{ga}\,\frac{\Delta t}{\Delta x} = \frac{k^2\,\sqrt{ga}}{\beta}\,\frac{\tau}{\xi} \le \sigma_0 \tag{9.36}$$

The following typical values have been used: $\sqrt{ga}\,k/r = 0.3$, $\beta/k^2\sqrt{ga} = 0.25\ 10^{-4}$, $F = 0.05$. The data are collected in table 9.2.

Table 9.2. Data for accuracy and amount of work for various methods

time discr.	grid	a	m	b	n	c	w
1. phase gravity waves (f = 0)							
LF	C	0.625	2	0.16	2	1	33
fw. Euler	unstable						
RK4	C	0.625	2	0.0068	4	2.8	150
RK4	FEM	0.011	4	0.0068	4	2.8	$1200 + 6n_x$
CN	C	0.625	2	0.05	2	-	
Euler s.i.	C	0.625	2	0.04	2	2	33
leap-frog s.i.1	C	= LF					
leap-frog s.i.2	C	0.625	2	0.28	2	-	$33 + 2n_x$
ADI	C	0.625	2	0.08	2	-	75
fractional step	E	0.069	2	0.08	2	-	$130 + 2n_x$
2. phase vorticity waves							
LF	C	0.625	2	0.138	2	0.035	33
fw. Euler	unstable						
RK4	C	0.625	2	$0.63\ 10^{-3}$	4	0.1	150
CN	C	0.625	2	0.069	2	-	
Euler s.i.	C	0.625	2	0.287	2	0.07	33
leap-frog s.i.1	C	= LF					
leap-frog s.i.2	C	= LF					$33 + 2n_x$
ADI	C	0.625	2	0.069	2	-	75
fractional step	E	"exact"				-	$130 + 2n_x$
3. damping flood waves							
LF	C	unstable					
fw. Euler	C	0.36	2	0.47	2		33
RK4	C	0.36	2	0.0022	4	0.59	150
RK4	FEM	0.045	4	0.0022	4	0.59	$1200 + 6n_x$
CN	C	0.36	2	0.07	3	-	
Euler s.i.	C	0.36	2	0.47	2	0.42	33
leap-frog s.i.1	C	0.36	2	0.57	3	0.21	33
leap-frog s.i.2	C	0.36	2	0.16	2	-	$33 + 2n_x$
ADI	C	0.36	2	0.019	3	-	75
fractional step	E	0.36	2	0.07	3	-	$130 + 2n_x$

time discr.	grid	a	m	b	n	c	w
3. phase Rossby waves							
LF	C	1.19	2	0.04	2	$0.18 \ 10^{-4}$	33
fw. Euler	unstable						
RK4	C	1.19	2	$0.39 \ 10^{-3}$	4	$0.50 \ 10^{-4}$	150
Euler s.i.	C	1.19	2	0.029	1	$0.35 \ 10^{-4}$	33
CN	C	1.19	2	0.06	2	-	
leap-frog s.i.1	C	1.19	2	0.04	2	$0.18 \ 10^{-4}$	33
leap-frog s.i.2	C	1.19	2	0.49	2	-	$33 + 2n_x$
ADI	C	1.19	2	0.06	2	-	75
fractional step	E	0.33	2	0.04	2	-	$130 + 2n_x$

Figs. 9.13 and 9.14 give some results for gravity waves. The figures are for an error level of 1 %. Only the case $f = 0$ has been considered, in which only phase errors occur (fig. 9.4). For most of the methods, the results will not change very much with nonzero f. On the horizontal axis of fig. 9.13, the number of grid points per wave length is given. However, please note that this *not* the number needed for a certain accuracy, which is determined in the optimization procedure and therefore not set by yourself. It is the number needed for representation of geometry and topography. If it is small, the algorithm will select a larger number n_x for accuracy and consequently the constraint is not active. This gives horizontal lines in the figure. At larger n_x, the constraint comes into action: the number of grid points is now larger than needed for accuracy of wave propagation and the amount of work increases quickly. In either case, stability may act as another constraint. Short lines on the left of the figure indicate values that would have been obtained without stability limits. For implicit methods there is no such limit.

Fig. 9.13 shows that without constraints there is little difference between leap-frog, ADI (both second-order methods) and even RK4 (4-th order), all on the same C-grid. The 4-th order time integration by itself does not lead to a more efficient method. If you combine it with a spatial discretization that is 4-th order accurate as well (e.g. finite elements on a regular grid), a significant improvement is found. However, if the grid size is constrained to larger n_x (right-hand side of the figure), all explicit methods are obliged to reduce their time steps for stability. The implicit ADI method does not need this and therefore emerges soon as the most efficient method.

You could wonder if other (semi-)implicit methods share this property. Fig. 9.14 compares the second semi-implicit leap-frog method and the fractional-step method with ADI. The fractional-step method is somewhat more efficient at low resolution due to the better accuracy of the E-grid for this case. Beware, however, of the amplitude error shown in fig. 9.3. It turns out that the curves for both methods rise very quickly at high resolution, but for a different reason: the operation count includes a term proportional to n_x. Using direct solution methods, the conclusion is that ADI is overall the most efficient method for gravity waves. It is not certain how relations would be if iterative solution techniques would be used instead.

Fig. 9.15 gives corresponding results for vorticity waves. It should be noted that the fractional step method is almost exact due to its characteristic treatment of advection,

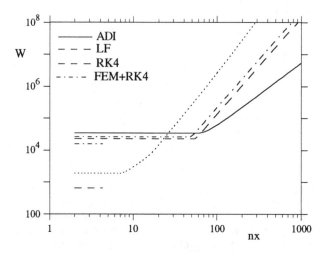

Fig. 9.13. Amount of work needed for an accuracy of 1 % in the phase of gravity waves, for various numerical methods, as a function of resolution set by representation of topography

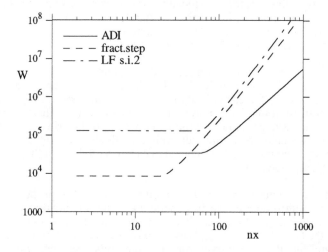

Fig. 9.14. Continuation of fig. 9.13

but it has a relatively high cost. Among the remaining methods, ADI is by far the most efficient, the reason being again that it is not restricted by stability conditions. Fourth-order time integration (RK4) does not pay for a 2nd order space discretization. Not shown is the Euler semi-implicit method; it would be slightly worse than leap-frog.

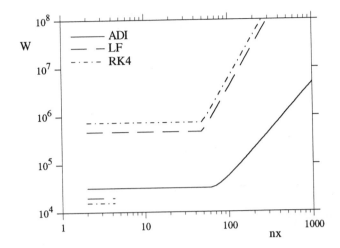

Fig. 9.15. Amount of work needed for an accuracy of 1 % in the phase of vorticity waves, for various numerical methods, as a function of resolution set by representation of topography

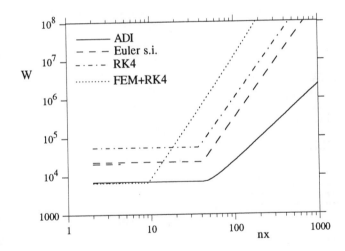

Fig. 9.16. Amount of work needed for an accuracy of 1 % in damping of flood waves, for various numerical methods, as a function of resolution set by representation of topography

For flood waves, fig. 9.16 shows again that there is relatively little difference at low resolution. ADI performs very well again due to its 3-rd order error: it reaches the same level of efficiency as a fully 4th order method (finite-elements with RK4). At higher resolution, the explicit methods are restricted by stability. The Euler method and its semi-implicit variant give reasonable results here, mainly because of their low cost. Note, however that the forward Euler method has stability problems if friction is weak.

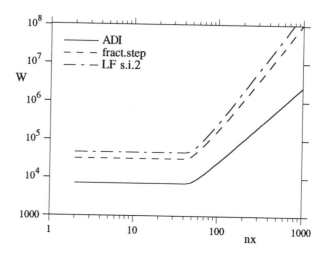

Fig. 9.17. Continuation of fig. 9.16

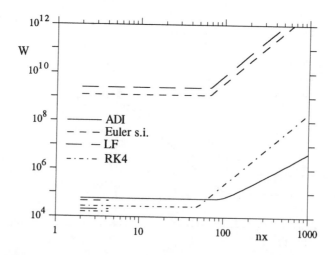

Fig. 9.18. Amount of work needed for an accuracy of 1 % in the phase of Rossby waves, for various numerical methods, as a function of resolution set by representation of topography

Some of the implicit methods are compared in fig. 9.17. Both semi-implicit leap-frog and fractional-step methods are less efficient than ADI due to their higher operation counts.

For Rossby waves, only the results for phase errors are given, although several methods display significant amplitude errors (fig. 9.8). In fig. 9.18 you see that there are very great differences in efficiency. The leap-frog and semi-implicit Euler methods are very expensive, mainly because of strict stability limits (remember that these are always for the fastest waves, that is gravity waves). The RK4 method is less restricted due to its high accuracy. The ADI method is about at the same level and again gets more efficient than the other methods at high resolution. Among the implicit methods (fig. 9.19), the fractional-step method is comparable to ADI but note its amplitude error. The second semi-implicit leap-frog method is less efficient and has an amplitude error as well.

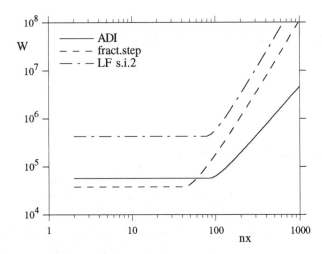

Fig. 9.19. Continuation of fig. 9.18

The overall conclusion appears to be that the ADI method is an efficient general-purpose method. Without resolution restrictions it may not be quite the best method but at least one of the better. At high resolution, it performs better than any of the other methods discussed here, due to its non-iterative solution method with a low operation count, combined with unrestricted stability. Some remarks are, however, in order. First of all, a specific choice has been made in these comparisons for some parameters in the stability limits and for the accuracy ($\varepsilon = 1$ %). Other values may lead to slightly different conclusions, but it seems improbable that the general tendency will be very much different. Secondly, these comparisons are based on estimated (not measured) data, particularly regarding the operation counts. Refined versions of the ADI method (such as those by Stelling, 1983) may have much higher operation counts. Moreover, it has been noted by Wilders et al (1988) and Stelling et al (1986) that there may be accuracy problems with the ADI method in case of poorly resolved topographic features. Yet, the conclusion that ADI-type methods are a good compromise between accuracy and computational effort seems to be well-supported. It agrees with the fact that many general-purpose operational codes for shallow-water problems are based on such methods.

Chapter 10

Numerical treatment of boundary conditions

10.1. Need for numerical boundary conditions

In chapter 5, you have seen the number and type of boundary conditions needed to get a well-posed problem. In the case of discretized equations (finite-difference, finite-element, spectral, etc.) the same boundary conditions are applied, but you need something additional because the number of boundary conditions is usually less than the number of unknowns at the boundaries. You could try to supplement this by using some of the discrete equations, but these usually cannot be applied right at the boundaries. This is most clearly seen for finite-difference equations, which involve some stencil of grid points around the central point. If applied at a boundary, you would need values from grid points located outside the region, which is impossible. Various approaches are possible to get additional equations on the boundary:

(i) apply some kind of extrapolation from the inner region for those variables that are not specified by boundary conditions;
(ii) introduce additional grid points outside the region such that the finite-difference equations can be used up to and including the boundary; however, this requires some way to define values at the added grid points (e.g. by symmetry arguments);
(iii) apply adapted finite-difference equations at the boundary, which use only available grid points, and consequently have an asymmetric grid stencil.

In all cases, you could say that additional "numerical boundary conditions" are constructed such that all variables can be computed at the boundaries. For example, in case (i) linear extrapolation of, say, the water level h from the interior to the boundary implies $\partial^2 h/\partial n^2 = 0$ (n is the normal direction) which is an additional assumption or numerical boundary condition. However, these are not additional boundary conditions in a physical sense (which would result in an overspecified problem) but just relations derived from information already available in the problem specification (i.e. differential equations and boundary conditions), possibly allowing some numerical error.

You will see in this chapter that there are good reasons to derive additional boundary equations by method (iii) above. This does not mean that the other methods do not work, but a theoretical justification for the third method is somewhat more evident.

10.2. Examples of boundary schemes

The finite-element method (see section 6.4) includes a systematic way to derive additional boundary relations. Physical boundary conditions are consistently taken into account. If variables are specified on the boundary (Dirichlet-type boundary conditions), these conditions are also applied in the discretized system. If gradients are given (Neumann-type conditions), these are taken into account as boundary integrals occurring in the weighted-residual formalism. For all unknowns not specified on the boundary (including those for which you have a Neumann condition), a weighted-residual equation is generated in the same way as for internal nodes. These equations for boundary nodes have the appearance of asymmetric finite-difference stencils.

As an example, we show the result for a triangular grid of uniform layout (fig. 6.9). Assume that the boundary $x = 0$ is in the middle of that figure. Suppose the water level h is not specified on the boundary. The shaded elements on the right-hand side of the boundary then give the following approximation of the continuity equation in linearized form (i.e. with constant advection velocities U, V and depth a):

$$M_2^* \frac{dh}{dt} + U D_{1x}^* h + V D_{1y}^* h + a D_{1x}^* u + a D_{1y}^* v = 0 \tag{10.1}$$

where an average has been taken between the two possible choices of element layout shown in fig. 6.9. The grid stencils marked with asterisks are variants of their counterparts in (6.21); they are given symbolically as

$$M_2^* = \frac{1}{12} \begin{bmatrix} 1 & 1 \\ 6 & 2 \\ 1 & 1 \end{bmatrix}$$

$$D_{1x}^* = \frac{1}{6 \, \Delta x} \begin{bmatrix} -1 & 1 \\ -4 & 4 \\ -1 & 1 \end{bmatrix}$$

$$D_{1y}^* = \frac{1}{6 \, \Delta y} \begin{bmatrix} 2 & 1 \\ 0 & 0 \\ -2 & -1 \end{bmatrix}$$

Similar expressions can be derived for the momentum equations, if either u or v is not specified on the boundary. Comparing (10.1) with (6.21), you notice that the mass and y-derivative stencils have just been split into right and left halves, of which the right one has been retained. The x-derivatives show the same y-averaging as (6.23) but necessarily involve forward differences in x-direction. The order of approximation is therefore less than for interior points, but this need not be a major problem as discussed in section 10.5.1. A similar reasoning applies to the spectral method in its weighted-residual form (section 6.7).

For finite-difference methods, you can also set up one-sided approximations at boundaries but there is no systematic procedure to do so. You may have to choose more or less heuristically how to approximate certain terms at a boundary, particularly on staggered grids.

Consider first the Arakawa C-grid (fig.10.1a) and suppose that there is a closed boundary at $x = 0$ (other cases are obviously similar). Then it is straightforward to arrange the grid as indicated such that u -points are on the boundary. The boundary condition of no normal flow is then easily implemented by fixing those u -values to zero. Things get more complicated in case of an open boundary where either 1 or 2 boundary conditions should be imposed depending on whether you have outflow or inflow. In the former case, if the water level h is specified, you arrange the grid to have h -points on the boundary (fig. 10.1b); a second condition on v is then possible. If the normal velocity u is specified, u -points are chosen to be on the boundary (fig. 10.1c). However, in this case, it is difficult to specify a second boundary condition at inflow. In fig. 10.1c, $v =0$ could be imposed only at a location one grid interval from the boundary. It would be even more complicated to impose a condition on the vorticity. Therefore, it is clear that you will get inaccuracies near the boundary. On the other hand, no "numerical conditions" are needed as the remaining variables are not even defined on the boundary and therefore need not be computed.

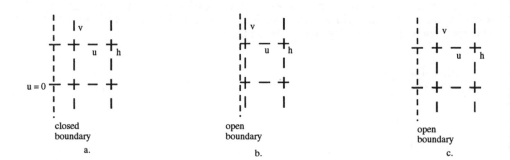

Fig. 10.1. Boundary arrangements in a C grid

On the E-grid (fig. 10.2), a boundary will always contain both (u,v)-points and h -points, so it is possible to impose all types of boundary conditions. Now, however, additional relations are needed. For example, on a closed boundary only $u = 0$ is given and you can use asymmetric finite-difference equations for h and v:

$$\frac{dh_{k,j}}{dt} + \frac{1}{\Delta x}\,(u_{k+1,j}M_y a_{k+1,j} - a_{k,j}M_y u_{k,j}) + D_{0y}(M_y a\, v) = 0 \qquad (10.2)$$

$$\frac{dv_{k,j}}{dt} + \frac{u}{\Delta x}(M_y v_{k+1,j} - v_{k,j}) + \nu M_y D_{0y} v + f u - \frac{g}{\Delta x}(h_{k+1,j} - M_y h_{k,j}) + rv = 0 \quad (10.3)$$

with suitable time-differencing (chapter 8). The averaging operator M_y is as defined in section 6.2. Similar procedures are possible for open boundaries.

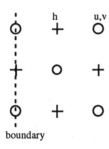

Fig. 10.2. Boundary arrangement on an E-grid

On the A-grid, the situation is very similar to the E-grid with the difference that additional "numerical conditions" are now needed at all grid points on the boundary. On the other hand, averages are not needed as all variables are defined in all grid points.

In the judgement of such approximations, two properties have to be discussed: stability and accuracy. The former dominates the literature on this subject because it turns out that even relatively obvious choices for boundary conditions may destabilize an otherwise stable method. Stability is discussed in sections 10.3 and 10.4. Much less is known about the influence of boundary treatment on accuracy. Some information is provided in section 10.5.

10.3. Stability analysis by the energy method

In section 5.2, the energy method has been used to analyse the type of boundary conditions needed to get a mathematically well-posed problem. The same method can be used to study the stability of finite-difference approximations. You will see that there is a very natural connection in some cases. However, the energy method cannot always be used and it must be supplemented by the normal-mode method discussed in section 10.4.

Consider as an example the Crank-Nicolson method applied to the linearized shallow-water equations (4.3) on the A-grid in a region with one boundary at $x = 0$ and homogeneous boundary conditions. It can be shown that a method which is stable in this case will generally also be stable for nonlinear equations with more boundaries and inhomogeneous boundary conditions (Strikwerda, 1989). However, there are some cases in which you get an interaction between two opposing boundaries (e.g. Beam et al, 1982, Trefethen, 1985).

The finite-difference equations can be written as

$$\mathbf{v}^{n+1} - \mathbf{v}^n + \tfrac{1}{2} \Delta t \, (AD_{0x} + BD_{0y} + C)(\mathbf{v}^{n+1} + \mathbf{v}^n) = 0 \tag{10.4}$$

at grid points k,j $(k > 0)$ and time steps n, and

$$\mathbf{v}^{n+1} - \mathbf{v}^n + \tfrac{1}{2} \Delta t \, (AD_{+x} + BD_{0y} + C)(\mathbf{v}^{n+1} + \mathbf{v}^n) = 0 \tag{10.5}$$

at the boundary $(k = 0)$ for those variables not specified by boundary conditions according to section 5.2. D_{+x} is a forward difference. For the energy argument, the system should be made symmetric as discussed in section 5.2. This result is that (10.4) and (10.5) are applied to (5.7). Omitting the primes, you can now suppose that the matrices A and B are symmetric. Premultiply (10.4) by $(\mathbf{v}^{n+1} + \mathbf{v}^n)^T$

$$\left(\mathbf{v}^T\mathbf{v}\right)^{n+1} - \left(\mathbf{v}^T\mathbf{v}\right)^n + 2\,\Delta t \,\overline{\mathbf{v}}^T(AD_{0x} + BD_{0y} + C)\,\overline{\mathbf{v}} = 0 \tag{10.6}$$

where $\overline{\mathbf{v}} = (\mathbf{v}^{n+1} + \mathbf{v}^n)/2$. A corresponding expression for $k = 0$ follows from (10.4). Numerical integration of (10.5) according to the trapezoidal rule (i.e. with a factor 1/2 in front of the boundary terms) gives a discrete energy equation. Concentrating on the x-differences, the integration gives

$$\left(\mathbf{v}_0^T A\mathbf{v}_1 - \mathbf{v}_0^T A\mathbf{v}_0\right) + \ldots + \left(\mathbf{v}_k^T A\mathbf{v}_{k+1} - \mathbf{v}_k^T A\mathbf{v}_{k-1}\right) + \left(\mathbf{v}_{k+1}^T A\mathbf{v}_{k+2} - \mathbf{v}_{k+1}^T A\mathbf{v}_k\right) + \ldots$$

Due to the symmetry of A, $\mathbf{v}_k^T A\mathbf{v}_{k+1} = \mathbf{v}_{k+1}^T A\mathbf{v}_k$ and the terms cancel at interior points. The only terms left are those on the boundary. Similarly, in y-direction all terms cancel because B is also symmetric. You could object that (10.5) is not used for all variables but only for those not specified by boundary conditions. Yet, (10.5) is used for all variables in the integration of (10.6). However, it is multiplied by the corresponding variable at the boundary. If this variable is specified at the boundary, it will be zero in this analysis as the boundary conditions have been assumed homogeneous. The result of the integration is therefore (after multiplication by Δx)

$$E^{n+1} - E^n + 2\,\Delta x\,\Delta t \sum_{k,j}{}' \overline{\mathbf{v}}^T C\overline{\mathbf{v}} - \Delta t\,\overline{\mathbf{v}}_0^T A\overline{\mathbf{v}}_0 = 0 \tag{10.7}$$

where $E^n = \Delta x \displaystyle\sum_{k,j}{}' \mathbf{v}_{k,j}^T \mathbf{v}_{k,j}$

and the prime on the summation sign indicates that values at $k = 0$ are multiplied by 1/2 (trapezoidal rule).

Now, (10.7) looks very much like the continuous energy balance (5.9) with integrals replaced by numerical integrations. The matrix C leads to damping; only bottom friction is involved as Coriolis terms drop out due their anti-symmetry. The last term in (10.7), when moved to the right-hand side, gives an energy growth due to boundary conditions (suppressing the j subscript):

$$\overline{v}_0^T A \overline{v}_0 = U(u_0^2 + v_0^2 + h_0^2) + 2\sqrt{ga}\, u_0 h_0 \qquad (10.8)$$

As you are considering only cases where u, v or h are prescribed on the boundary, the energy argument does not produce statements on more general boundary conditions such as (5.12) or (5.14). For homogeneous boundary conditions, you have the following special cases of the general argument shown in section 5.2:

Case 1. For a closed boundary, $U = 0$ and the (homogeneous) boundary condition will be $u_0 = 0$. Then the contribution (10.8) to the energy balance is zero, the total energy will decrease (due to the dissipative effect of C) and the system is stable.

Case 2. For inflow, $U > 0$. If $h_0 = 0$ and $v_0 = 0$ are specified as boundary conditions, (10.8) gives a positive result, corresponding to an energy growth. Stability is not assured then, but it is still possible due to the damping effect of C. This must be analyzed by the normal mode analysis (section 10.4). If, on inflow, u and v are specified, you get a similar result: stability is undecided.

Case 3. On outflow, $U < 0$ and only one boundary condition can be specified. If this is either $u = 0$ or $h = 0$, the remaining part of (10.8) will be negative and stability is proved.

For the Crank-Nicolson method on the C-grid, you can follow a similar procedure. Again, the symmetrized equations (5.7) are used. In each grid point where a variable is computed (e.g. a u-point), you multiply the difference equation (the $x$momentum equation) by this variable (averaged between the old and new time levels) and add the result for all grid points. It is then easy to check that contributions from all internal points cancel except those from the bottom friction terms. At the boundary, you are left with some terms (like those in (10.8)) that do not cancel.

Case 1. At a closed boundary at $x = 0$ ($k = 1$), u-points are on the boundary and the mean velocity U across the boundary must be zero. Then the "leftover" term in the energy balance is $h_1 u_0$. For a homogeneous boundary condition, $u_0 = 0$ so the boundary contribution vanishes and the system is stable.

Case 2. At an open boundary where the water level is specified, h-points will be on the boundary. Then the "leftover" terms are

$$U(u_1^2 + v_0^2 + h_0^2) + \sqrt{ga}\, h_0 u_1$$

(the j-subscript is again suppressed). On inflow ($U > 0$), homogeneous boundary conditions $h_0 = 0$, $v_0 = 0$ could be specified, but this would still allow energy growth from the first term. You cannot decide whether or not this will be controlled by bottom friction. In this case, therefore, you do not get a stability proof from the energy method. Again, this does not imply that the method is unstable; you just need a different method to decide.

Case 3. In case of outflow ($U < 0$), only one condition is given, usually on the water-level, and you get an energy growth term

$$U (u_1^2 + v_0^2)$$

which is now negative, so you have stability. If you specify the velocity instead, the results are about the same: stability for outflow, undecided for inflow.

On the E-grid, a Crank-Nicolson variant of (6.6) can be used, together with the boundary treatment shown in section 10.2. Again, in each grid point, you multiply the difference equation by the variable to which it applies, averaged in time. Next, integration by the trapezoidal rule gives an energy equation. It is straightforward to see that all differences in y-direction cancel. Taking only x-differences into account, you obtain for rows having a (u,v) point on the boundary (suprressing the j-subscript)

$$U\{u_k M_y(u_{k+1} - u_{k-1}) + ... + u_0(M_y u_1 - u_0)\} +$$
$$+\sqrt{ga}\{u_k(h_{k+1} - h_{k-1}) + ... + u_0(h_1 - M_y h_0)\}+$$
$$+ \sqrt{ga}\{h_{k+1}(u_{k+2} - u_k) + ... + h_1(u_2 - u_0)\}$$

and for rows having an h-point on the boundary

$$U\{u_{k-1} M_y(u_k - u_{k-2}) + ... + u_1 M_y(u_2 - u_0)\} +$$
$$+\sqrt{ga}\{u_{k-1}(h_k - h_{k-2}) + ... + u_1(h_2 - h_0)\}+$$
$$+ \sqrt{ga}\{h_k(u_{k+1} - u_{-1}) + ... + h_0(u_1 - M_y u_0)\}$$

Similar expressions for v should be included. If these expressions are added for all rows of grid points, all terms cancel except an energy growth term similar to (10.8)

$$\sum_j \{U(u_{0,j+1}^2 + v_{0,j+1}^2 + h_{0,j}^2) + \sqrt{ga}\, u_{0,j+1}(h_{0,j} + h_{0,j+2})\} \qquad (10.9)$$

The same cases as before can be discerned.

Case 1: closed boundary, $U = 0$, $u_{0,j+1} = 0$. The result of (10.9) is zero, so there is stability.

Case 2: open boundary with inflow. Boundary conditions $u_{0,j+1} = 0$ and $v_{0,j+1} = 0$ allow possible energy growth: stability is undecided. If the boundary conditions are $h_{0,j} = 0$, $v_{0,j+1} = 0$, the same conclusion is drawn.

Case 3: open boundary with outflow. If either $u_{0,j+1} = 0$ or $h_{0,j} = 0$, the result of (10.9) is negative, so you have stability.

As a conclusion, for the Crank Nicolson method on either the A, C or E grids, the stability behaviour is the same as in the continuous case. Unfortunately, this cannot be generally proved for other time discretization methods shown in ch. 8, especially the mixed ones.

The standard leap-frog method would get unstable by inclusion of Coriolis and bottom friction terms. In the first semi-implicit leap-frog method (8.19), the latter are treated implicitly. If you do the same for the Coriolis terms, you get another semi-implicit variant

$$\mathbf{v}^{n+1} - \mathbf{v}^{n-1} + 2 \, \Delta t \, (AD_{0x} + BD_{0y})\mathbf{v}^n + \Delta t \, C(\mathbf{v}^{n+1} + \mathbf{v}^{n-1}) = 0 \qquad (10.10)$$

with boundary conditions based on this same equation with forward instead of central x-differences. You could try to do the same type of energy analysis as in the previous section, that is multiply by $\mathbf{v}^{n+1} + \mathbf{v}^{n-1}$. However, this gives a contribution

$$\Delta t \, (\mathbf{v}^{n+1} + \mathbf{v}^{n-1})^T (AD_{0x} + BD_{0y})\mathbf{v}^n$$

which cannot be reduced to boundary values only as in (10.7) even if A and B are symmetric. You could have a situation with $\mathbf{v}^n$ behaving like $(-1)^n$ (the leap-frog method normally has a tendency to develop such oscillations in time) and the boundary contribution would be of the wrong sign. The energy method then does not show stability and you would have to use another method of analysis.

10.4. Normal mode analysis

A very powerful theory of stability for initial-boundary value problems was given by Gustafsson, Kreiss & Sundström (1972). The analysis, known as the GKS method, can be considered an extension of the well-known Von Neumann stability analysis, which is valid for pure initial-value problems without boundaries and decides whether any harmonic (Fourier) component of the discrete solution can grow exponentially. In the GKS or normal-mode analysis, this is extended to non-harmonic normal modes generated by the numerical treatment of the boundary conditions. If normal modes with amplitudes increasing in time can be supported by homogeneous boundary conditions (that is: with no energy input), the system is unstable. A very clear account of the theory is given by Strikwerda (1989). A more physical interpretation is presented by Trefethen (1982, 1983). Unfortunately, although the analysis is straightforward in principle, it turns out to be very complicated in practice. Therefore, here only some examples are given which explain how the method works.

10.4.1. A one-dimensional example

Consider the linearized and symmetrized one-dimensional shallow-water equations without bottom friction terms:

$$\frac{\partial \mathbf{v}}{\partial t} + A \frac{\partial \mathbf{v}}{\partial x} = 0 \tag{10.11}$$

with

$$\mathbf{v} = \begin{pmatrix} u \\ \sqrt{g/a}\, h \end{pmatrix} \qquad A = \begin{pmatrix} U & \sqrt{ga} \\ \sqrt{ga} & U \end{pmatrix}$$

Eigenvalues of A are

$$c_{1,2} = U \pm \sqrt{ga}$$

with corresponding eigenvectors

$$\mathbf{V}_1 = \begin{pmatrix} 1 \\ 1 \end{pmatrix} \qquad \mathbf{V}_2 = \begin{pmatrix} 1 \\ -1 \end{pmatrix}$$

Suppose that you solve (10.11) by means of the Crank-Nicolson method, i.e.

$$\mathbf{v}^{n+1} - \mathbf{v}^n + \tfrac{1}{2} \Delta t\, A\, D_{0x}(\mathbf{v}^{n+1} + \mathbf{v}^n) = 0 \tag{10.12}$$

on a region $x \geq 0$ with $h_0^n = f(t_n)$ specified as a boundary condition. It is sufficient to consider a homogeneous condition, $f(t) = 0$. As a second, numerical boundary condition, you can use either of the two equations with one-sided differences

$$\mathbf{v}^{n+1} - \mathbf{v}^n + \tfrac{1}{2} \Delta t\, A\, D_{+x}(\mathbf{v}^{n+1} + \mathbf{v}^n) = 0 \quad \text{at} \quad x = 0 \tag{10.13}$$

For the normal mode analysis, solutions of the homogeneous equations are studied which have the form

$$\mathbf{v}_j^n = \mathbf{V}\, z^n \kappa^j \tag{10.14}$$

If $\kappa = e^{i\xi}$ i.e. $|\kappa| = 1$, this is just the case of Von Neumann's stability analysis and z would then be the amplification factor. The method is assumed to be stable in the Von Neumann sense for a pure initial-value problem, as otherwise the GKS analysis does not make sense. For the Crank-Nicolson method, you have seen this to be true (section 8.3). In the normal-mode analysis, general complex values of κ are allowed. Inserting (10.14) into (10.12), you get homogeneous algebraic equations for the amplitude vector $\mathbf{V}$:

$$\left((z-1)I + \frac{\Delta t}{4\Delta x} (z+1)(\kappa - \frac{1}{\kappa})A \right) V = 0$$

where I is the unit matrix. This shows that V is an eigenvector of A. The eigenvalues satisfy

$$\frac{z-1}{z+1} = -\frac{1}{4}\sigma_{1,2}(\kappa - \frac{1}{\kappa}) \qquad (10.15)$$

where $\sigma_{1,2} = c_{1,2} \Delta t/\Delta x$ are the two Courant numbers. Eq.(10.15) is called the *resolvent equation*; for each z it gives possible values for κ. For each Courant number, there are two such values, so the general solution consists of four contributions, each of the form (10.14) with common z but different κ and corresponding eigenvector V. You can easily check that if a certain κ is a solution of (10.15), so is $(-1/\kappa)$. Consequently, solutions occur in pairs and in each pair, one has $|\kappa| > 1$ and the other $|\kappa| < 1$. The former gives a contribution to (10.14) that grows indefinitely for large j (or x) and is therefore not allowed: its amplitude should be zero. For each z, the remaining solution now consists of two components behaving well for large j:

$$v_j^n = z^n \{a_1 V_1 \kappa_1^j + a_2 V_2 \kappa_2^j\} \qquad (10.16)$$

with $|\kappa_{1,2}| < 1$. The two coefficients $a_{1,2}$ are determined by the boundary conditions. The homogeneous "physical" boundary condition $h = 0$ gives

$$a_1 V_{1,2} + a_2 V_{2,2} = 0 \qquad (10.17a)$$

($V_{1,2}$ is the second component of the first eigenvector etc.). If you use the second component (that is the continuity equation) of (10.13) as a second boundary condition, this gives after substitution of (10.16), isolating the second component of the vector equation and realizing that V is an eigenvector:

$$\frac{z-1}{z+1} (a_1 V_{1,2} + a_2 V_{2,2}) + \frac{1}{2}\sigma_1(\kappa_1-1)a_1 V_{1,2} + \frac{1}{2}\sigma_2(\kappa_2-1)a_2 V_{2,2} = 0 \qquad (10.17b)$$

The two homogeneous equations (10.17) determine the coefficients $a_{1,2}$. Generally, they will only have a zero solution which means that no growing solutions are possible and the numerical method is stable. However, if these equations allow a nontrivial growing solution, you see that an unstable component of the numerical solution can occur spontaneously. This happens if the determinant is zero; after some algebra you find this to be

$$\sigma_1(\kappa_1-1) - \sigma_2(\kappa_2-1) = 0 \qquad (10.18)$$

So the question is: can this happen for $|z| > 1$? For each σ and z, κ can be solved from (10.15):

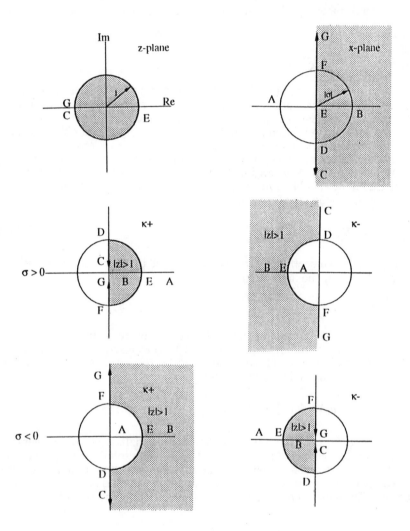

Fig. 10.3. Behaviour of z, x, κ in the complex plane (see text)

$$\kappa_{1,2} = \frac{x}{\sigma}\left(-1 \pm \sqrt{\left(\frac{\sigma}{x}\right)^2 + 1}\right) \qquad \left(x = 2\frac{z-1}{z+1}\right) \tag{10.19}$$

where you should select that root that has $|\kappa| < 1$. It is somewhat complicated to make this selection for complex valued variables. The situation is shown in fig. 10.3. The region outside the unit circle in the z-plane ($|z| > 1$) is mapped to the right half of the complex x-plane. If x is on the imaginary axis and $|x| < |\sigma|$, both roots $\kappa_{1,2}$ are on the unit circle. If x is on the imaginary axis but $|x| > |\sigma|$, one of the κ roots returns to the origin but the other goes to infinity. The consequence is that if $\sigma > 0$, you choose κ^+ as the root which has $|\kappa| < 1$ for $|z| > 1$. Conversely, for $\sigma < 0$, you choose the root κ^-.
Without loss of generality, you can assume that $\sigma_1 > 0$. You can now substitute the correct root from (10.19) into (10.18):

$$x\left(-1 + \sqrt{\left(\frac{\sigma_1}{x}\right)^2 + 1}\right) \cdot \sigma_1 - x\left(-1 - \sqrt{\left(\frac{\sigma_2}{x}\right)^2 + 1}\right) + \sigma_2 = 0$$

or, after some algebra

$$(\sigma_1 - \sigma_2)^2 = 0$$

which is clearly not possible. The conclusion is therefore that solutions to the homogeneous equations and boundary conditions with $|z| > 1$ are not possible and the system is stable.

The other possibility to handle the boundary is using the first component of (10.13) (the momentum equation) together with the boundary condition $h = 0$. Then instead of (10.17b) you get

$$\frac{z-1}{z+1}(a_1 V_{1,1} + a_2 V_{2,1}) + \frac{1}{2}\sigma_1(\kappa_1-1)a_1 V_{1,1} + \frac{1}{2}\sigma_2(\kappa_2-1)a_2 V_{2,1} = 0 \tag{10.20}$$

and the determinant condition becomes (using the eigenvectors given above)

$$2\frac{z-1}{z+1} + \sigma_1(\kappa_1-1) + \sigma_2(\kappa_2-1) = 0 \tag{10.21}$$

The same selection of roots can be used, so that this can be rewritten as

$$(\sigma_1 + \sigma_2)^2 = 0$$

Now, as $\sigma_{1,2} = (U \pm \sqrt{ga})\,\Delta t/\Delta x$ this *could* happen but only in the special case $U = 0$. On an open boundary, this will happen only occasionnally, e.g. during the turning of the tide. At such moments, an instability might set in, but is is unlikely to continue. As the previous method of using the continuity equation at the boundary does not have this difficulty, it is to be preferred.

10.4.2. A two-dimensional example

For the 2-d SWE, a similar normal-mode analysis can be done for one boundary at a time, say the one at $x = 0$. To reduce the problem to one spatial dimension, a Fourier transform is taken in y-direction (transformed variable k_2). A complication is that you now have three equations and more possible combinations of boundary conditions. As an example, consider the Crank-Nicolson method on an A-grid. This gives the following equation for the normal modes:

$$\left(\frac{z\text{-}1}{z\text{+}1} I + \frac{\Delta t}{4\Delta x} (\kappa - \frac{1}{\kappa})A + i \frac{\Delta t}{2\Delta y} B \sin \eta \right) V = 0 \tag{10.22}$$

with $\eta = k_2 \Delta y$. Unfortunately, the eigenvectors are no longer those of A and determining the values of κ and V for general z, and particularly determining which of them are inside the unit circle, is hardly possible by hand. You might manage to do it using computer algebra.

To get an impression at least, you could take the special case $\eta = 0$ (that is, waves propagating normal to the boundary), but this is, of course, not a complete analysis. The only conclusion to draw is that boundary conditions might already be unstable in that case. Elvius & Sundström (1973) took the same approach for their semi-implicit method. In this case, you have the eigenvalues of A with eigenvectors

$$c_1 = U \quad c_2 = U + \sqrt{ga} \quad c_3 = U - \sqrt{ga}$$

$$V_1 = \begin{pmatrix} 0 \\ 1 \\ 0 \end{pmatrix} \quad V_2 = \begin{pmatrix} 1 \\ 0 \\ \sqrt{a/g} \end{pmatrix} \quad V_3 = \begin{pmatrix} 1 \\ 0 \\ -\sqrt{a/g} \end{pmatrix}$$

The general solution of the Fourier-transformed finite-difference equations is

$$v_j = \sum_{k=1}^{3} a_k V_k \kappa_k^j \quad (|\kappa_k| < 1)$$

where κ is the solution of

$$2 \frac{z\text{-}1}{z\text{+}1} + \sigma_k(\kappa_k - 1/\kappa_k) = 0 \qquad k = 1,2,3 \tag{10.23}$$

The analysis of this equation is the same as in the 1-d case, so for any $\sigma > 0$, you take the + root and otherwise the - root.

For the case of *inflow*, you can take $u = 0$ and $v = 0$ as boundary conditions, together with the continuity equation (third component of the system of equations) as additional condition. Using forward x-differences at the boundary as in the 1-d case, you get the following determinant equation from the boundary conditions:

$$\begin{vmatrix} V_{11} & V_{21} & V_{31} \\ V_{12} & V_{22} & V_{32} \\ p_1 V_{13} & p_2 V_{23} & p_3 V_{33} \end{vmatrix} = 0 \quad (p_k = 2x + \sigma_k(\kappa_k - 1))$$

or

$$-2x - \sigma_3(\kappa_3 - 1) - 2x - \sigma_2(\kappa_2 - 1) = 0 \tag{10.24}$$

This is the same condition as (10.21) in the one-dimensional case, using the momentum equation at the boundary. You saw that it will only be satisfied at flow reversal across the boundary, so this approach may be expected to be stable. However, remember that this does not say anything about waves moving in other than normal directions. Similarly, if you use boundary conditions $h = 0$ and $v = 0$ on inflow, you again find the same result as in the one-dimensional case.

At outflow, you can specify only one boundary condition, say $h = 0$, complemented for example by the two momentum equations. The determinant equation for the boundary then becomes

$$\begin{vmatrix} V_{13} & V_{23} & V_{33} \\ p_1 V_{12} & p_2 V_{22} & p_3 V_{32} \\ p_1 V_{13} & p_2 V_{23} & p_3 V_{33} \end{vmatrix} = 0 \quad (p_k = 2x + \sigma_k(\kappa_k - 1))$$

This splits into two parts; one is identical to the case of inflow, which was found to be stable, and the other is related to the slow vorticity wave:

$$2x + \sigma_1(\kappa_1 - 1) = 0 \tag{10.25}$$

which using (10.23) can be rewritten as

$$\sigma_1(\kappa_1 - 1) = 0$$

This could be satisfied if $z = 1$. However, this very special case can be considered as a limit of non-admissible waves (with $|z| > 1$ but $|\kappa| > 1$, see fig. 10.3) and the theory then says that it will not lead to instability. For the case that $u = 0$ is prescribed at outflow, you get the same result.

10.5. Accuracy of boundary treatment

10.5.1. Order of accuracy

The suggestions made in section 10.2 for finite-difference equations at the boundaries, needed to supplement the physical boundary conditions, generally have a lower order of accuracy than the equations used at internal grid points. Eqs. (10.1)...(10.3) involve forward x-differences which are only first-order accurate in space. Whether or not this is acceptable is an accuracy question.

If the system of equations is *hyperbolic*, that is if lateral viscosity terms are disregarded, a theoretical result by Gustafsson (1975) states that boundary approximations may be one order lower than internal ones without destroying the order of convergence. Most of the techniques described in this book are second-order accurate in space, so Gustafsson's theory says that you can afford first-order accurate boundary treatment (like 10.1) and still have second-order *convergence*. This does not imply that the error due to boundary treatment is actually small at a certain grid size, but it will decrease quadratically with decreasing grid size. The FEM boundary approximation (10.1) is locally only first-order accurate. Therefore, it is likely that the 4th order accuracy of the method on a regular grid will be reduced to second order globally.

For a *parabolic* system of equations (which is what you get by including lateral viscosity terms), this is no longer true: in order to obtain second-order convergence you must approximate the equations at the boundary also with second-order accuracy. An example is given by Fletcher (1988). In (10.2),(10.3), this would mean that asymmetric 3-point x-differences should be taken. Stability of such approximations would have to be investigated along lines shown in the previous sections. Many practical applications have ignored this rule without obvious difficulties. This might be explained by noting that the viscosity terms, if taken into account, will usually have only a minor influence, so that the system remains "almost" hyperbolic. A theoretical justification is missing, however.

In general, very little is known about the influence of boundary treatment on accuracy. Only a few specific cases have been studied in literature, of which two examples are shown.

10.5.2. Oblique boundaries

On a rectangular grid, oblique boundaries have to be approximated by "zig-zag" grid-aligned boundaries (fig. 6.4.a). Imagine a closed boundary under 45° (fig. 10.4) in a C-grid. The condition of no normal flow implies that $u = 0$ and $v = 0$ at alternating boundary points and this might amount to the specification of these two boundary conditions simultaneously. For a "viscous" simulation, this is perfectly acceptable (although you might prefer a free-slip or partial-slip condition), but not in a non-viscous one. In the latter case, only the normal velocity component should vanish, so $u - v = 0$ with the grid-aligned components u and v both nonzero.

A full analysis of this case is difficult, but a partial analysis by Weare (1979) indicates that the problem is present only in ADI methods and then only for unsteady flows. He considers as a special case a very simple accelerating flow, described in the channel-aligned coordinate system (fig. 10.4) as

$$\frac{\partial u'}{\partial t} + g\frac{\partial h'}{\partial x'} = 0 \tag{10.26}$$

with $\quad \frac{\partial u'}{\partial x'} = 0 \quad$ and $\quad \frac{\partial h'}{\partial x'} = -I \quad$ fixed

The obvious solution is $u' = gIt$, $v' = 0$, independent of x and y.

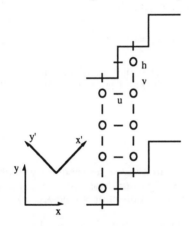

Fig. 10.4. Oblique boundaries in a C-grid

On the grid shown in fig. 10.4, a semi-discrete solution would be found from

$$\frac{\partial u_{k,j}}{\partial t} + g\,D_{0x}h_{k,j} = 0 \qquad j = 3,5,...J\text{-}2$$
$$\frac{\partial v_{k+1,j}}{\partial t} + g\,D_{0y}h_{k+1,j} = 0 \quad j = 4,6,...J\text{-}1 \tag{10.27}$$

with boundary conditions $u_{k,1} = u_{k,J} = 0$ and $v_{k+1,2} = v_{k+1,J+1} = 0$. Pressure would be determined from the continuity equation

$$D_{0x}u_{k+1,j} + D_{0y}v_{k+1,j} = 0 \quad j = 3,5,...J \tag{10.28}$$

or $\quad D_{0x}^2 h_{k+1,j} + D_{0y}^2 h_{k+1,j} = 0 \quad j = 5,...J\text{-}2$ (10.29)

Assume that there is a uniform flow with a pressure gradient such that

$$h_{k+1,j} = h_{k-1,j-2} - \Delta h$$ (10.30)

Then (10.29) becomes just

$$D_{0y}^2 h_{k+1,j} = 0 \quad j = 5,...J\text{-}2$$

with solution $h_{k+1,j} = A + Bj$ (*A* and *B* are integration constants to be determined). However, at the boundaries, (10.28) becomes with (10.30)

$$h_{k+1,5} - h_{k+1,3} = -\tfrac{1}{2}\Delta h, \quad h_{k+1,J} - h_{k+1,J-2} = -\tfrac{1}{2}\Delta h$$

which leaves *A* undetermined (this is not serious as only gradients of *h* are important) and $B = -\Delta h/4$. Finally, substituting this into (10.27) and realizing that $\Delta h = 2\sqrt{2}\, I\, \Delta x$, you get

$$\frac{\partial u_{k,j}}{\partial t} = \frac{\partial v_{k+1,j}}{\partial t} = \frac{gI}{\sqrt{2}}$$

which agrees with the exact solution. So in this case, the seemingly double boundary condition $u = v = 0$ does not give difficulties. This would be valid for any straightforward time discretization of the equations (e.g. Crank-Nicolson or Runge-Kutta).

Weare (1979) argues that an ADI solution does introduce difficulties here. If you take the ADI formulation (8.25), combine the two steps into one and specialize to the example of (10.27), you find the finite-difference equations

$$u^{n+1} - u^n + g\Delta t\, D_{0x} \bar{h} + \lambda\, D_{0x}D_{0y}(v^{n+1} - v^n) = 0$$

$$v^{n+1} - v^n + g\Delta t\, D_{0y} \bar{h} = 0$$ (10.31)

$$\bar{h} = \theta h^{n+1} + (1-\theta)h^n$$

with $\lambda = \theta^2 \Delta t^2 ga$. The latter term in (10.31a) is spurious and caused by the ADI approximation. If you now rotate the coordinate system and assume $v' \approx 0$, you have $u = v = u'/\sqrt{2}$ and (10.31a) gives

$$(u')^{n+1} - (u')^n + g\Delta t\, D_{0x'}\, \bar{h} + \lambda\, D_{0x}D_{0y}\{(u')^{n+1} - (u')^n\} = 0$$

Again, if you assume uniform flow, $D_{0x}u = - D_{0y}u$ and the equation can be considered an approximation of

$$\frac{\partial u'}{\partial t} - gI - \frac{1}{2} \lambda \frac{\partial^3 u'}{\partial t \, \partial y^2} = 0 \tag{10.32}$$

This has a solution $u' = t \, (\tilde{u} + gI)$ with

$$\tilde{u} - \frac{1}{2} \lambda \frac{\partial^2 \tilde{u}}{\partial y^2} = 0 \tag{10.33}$$

Compared with (10.26), you have added a kind of lateral viscosity of numerical origin, but with a very real effect. The double boundary condition $u = v = 0$ now *does* work as a no-slip boundary condition. Weare verified that the solution of (10.33) with no-slip boundary conditions agrees with his numerical results (fig. 10.5). Similar conclusions can be drawn if other terms, such as bottom friction, are included.

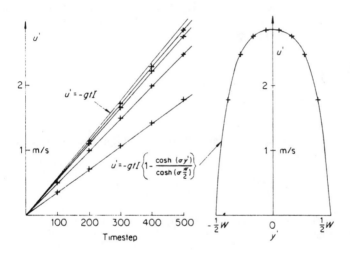

Fig. 10.5. Numerical results (+) for an accelerating flow in an oblique channel (from Weare, 1979).
Left: velocity as function of time at several locations.
Right: Velocity profile across channel. Drawn lines: theory.

It should be noted that if you are looking for a steady-state solution, it will be correct even if an ADI method is used. You can see this in (8.25) which reduces to the unsplit steady-state equation $(P + Q) \, v = 0$ as it should. The apparent viscosity noted by Weare therefore works only in unsteady flows. This is also obvious from (10.32). There does not seem to be a way of avoiding it, but you could try to find a set of boundary conditions implying no normal flow $v' = 0$ and free slip $\partial u'/\partial y' = 0$, which more or less reflect what you have in mind for the original "nonviscous" flow. According to Weare, however, it is not possible to do so without coupling the two ADI stages together, which would remove the advantage of the technique. This drawback of the ADI method has led several groups to consider unsplit, fully implicit or semi-implicit methods (Benqué et al, 1982, Wilders et al, 1988).

10.5.3. Outflow boundaries

As described in ch. 5, one physical boundary condition is needed on outflow boundaries (for the non-viscous case). Usually, this will be a specification of the downstream water level. On a non-staggered grid, e.g. the A-grid, this has to be supplemented by two additional numerical relations. It turns out that these often give rise to unphysical "wiggles" in the computed flow field. An analysis of this phenomenon has been given by Vreugdenhil & Booij (1985).

It is sufficient to consider a steady flow. Linearizing about a basic steady state, you obtain (4.1). As the "wiggles" are usually not obviously present in the water level, you can suppose for simplicity that the flow has a rigid lid, i.e. the water level is assumed fixed (and determined by the physical downstream boundary condition). The remaining disturbance of the flow field is then described by

$$U \frac{\partial u}{\partial x} + V \frac{\partial u}{\partial y} + ru = 0$$
$$U \frac{\partial v}{\partial x} + V \frac{\partial v}{\partial y} + rv = 0$$

$$(10.34)$$

where the Coriolis acceleration has been omitted for simplicity. The two equations are uncoupled; their analytic solutions are

$$u(x,y) = u_0(y - \frac{V}{U} x) \, exp(- \frac{rx}{U})$$

and similarly for v. The functions u_0, v_0 are fixed by the two required conditions at the upstream (inflow) boundary.

The numerical equivalent of (10.34a) could be (the way of time-differencing is of no importance)

$$U D_{0x}u + V D_{0y}u + ru = 0$$

$$(10.35)$$

This has solutions of the form $u_{k,j} = \kappa^k s^j$ (note the similarity with normal-mode analysis) provided κ and s satisfy

$$\kappa - \kappa^{-1} + \frac{V}{U}(s - s^{-1}) + 2r\Delta x/U = 0 \tag{10.36}$$

Suppose that the upstream boundary condition can be written as $u_{0,j} = u_0 s^j$ with some specified value of s. Then the two possible values for κ are

$$\kappa_{1,2} = -\varepsilon \pm \sqrt{1 + \varepsilon^2} \approx -\varepsilon \pm 1 \tag{10.37}$$

where $\varepsilon = \dfrac{2\, r\Delta x + V(s - s^{-1})}{2U}$

which is assumed to be small. The + root in (10.37) is the physical one; the - root oscillates in space and is spurious. The general numerical solution looks like

$$u_{k,j} = s^{\,j}(A\,\kappa_1^k + B\,\kappa_2^k) \tag{10.38}$$

in which the constants A and B are determined by the boundary conditions. At $x = 0$ (inflow)

$$A + B = u_0 \tag{10.39}$$

For the downstream "numerical boundary condition", several possibilities can be studied.

Take, e.g. a simple extrapolation $u_{n,j} = u_{n-1,j}$. From (10.38)

$$A\,(\kappa_1^n - \kappa_1^{n-1}) + B\,(\kappa_2^n - \kappa_2^{n-1}) = 0 \tag{10.40}$$

Solving for A and B and inserting (10.37), you find

$$A \approx u_0, \qquad B \approx (-1)^n \frac{\varepsilon}{2} u_0$$

The first component is therefore close to the correct solution. The oscillating component B is small (though first-order in Δx) and you get only weak wiggles.

Another possible way of handling the boundary conditions is adding a "virtual" row of grid point $(n+1)$ and assuming some sort of symmetry: $u_{n+1,j} = u_{n-1,j}$. In this case, (10.40) will be replaced by

$$A\,(\kappa_1^{n+1} - \kappa_1^{n-1}) + B\,(\kappa_2^{n+1} - \kappa_2^{n-1}) = 0 \tag{10.41}$$

Together with (10.39), you find now (assuming n uneven for simplicity) $A \approx B \approx u_0/2$, so the unphysical component is as strong as the physical one and you have an $O(1)$ error.

Thirdly, you could extrapolate linearly by assuming $u_{n+1,j} - 2u_{n,j} + u_{n-1,j} = 0$. In exactly the same way you can derive

$$A \approx u_0, \qquad B \approx (-1)^n \frac{\varepsilon^2}{4} u_0$$

which is second-order accurate and the spurious component is very small.

These results were verified numerically. Fig. 10.6 shows numerical results by Vreugdenhil & Booij (1985) using the first two methods of extrapolation. The strong, $O(1)$ oscillations in the case of extrapolation over two grid intervals are obvious.

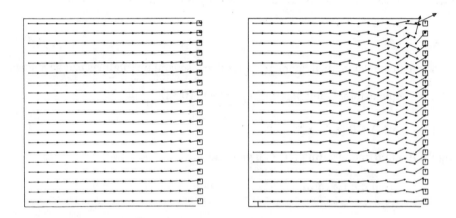

Fig. 10.6. Computed steady-flow pattern with different numerical outflow conditions. Left: extrapolation over one interval (10.40), right: symmetry (10.41)

Chapter 11

Three-dimensional shallow-water flow

11.1. Why 3-d ?

Most of this book is about 2-d shallow-water flows, obtained either by integrating over depth or by splitting into vertical modes. This approach works remarkably well in many cases, even when you would expect significant 3-d effects, such as in wind-driven flows. However, there are many other cases where the 2-d model is not good enough but where the shallow-water approximation can still be made. In such cases, you will have to solve the 3-d SWE (or, what amounts to the same thing, boundary-layer equations). Some of the reasons may be the following.

(i) The approximations made in chapter 2 for the 2-d SWE may not be valid. This applies in particular to bottom friction (section 2.7) and lateral momentum exchange (section 2.8). For example, if you have relatively short-period waves, the quasi-steady formulation of bottom stress may break down. This matters, of course, only if bottom friction is important at all, which you can estimate using section 4.4. Johns & Oguz (1987) give an example of a storm surge where it turns out that bottom roughness in the deeper areas is relatively unimportant.

(ii) You may want to know the 3-d flow structure, e.g. to compute spreading of pollutants in water. Spreading by velocity differences over the vertical (differential advection or shear dispersion) is an important mechanism. This is amply illustrated, e.g., in the book by Fischer et al (1979). They show that in some cases reasonable estimates can be made by assuming some velocity profile over depth (say, logarithmic or polynomial), but in more complicated cases this does not work. This is particularly so in near-field problems, such as cooling water outfalls.

(iii) Density differences induce a 3-d flow structure. This is by far the most important reason for the 2-d SWE to be ineffective for density-driven flows, unless they are very well mixed over depth.

For these and similar reasons, 3-d shallow-water computer models based on numerical solution of (2.16) have been set up since about 1970. This is not a historical survey but a few of the pioneers should be mentioned without any claim of completeness. The field seems to have been opened up for oceanographic applications by Bryan (1969), albeit with no free-surface motions (rigid lid approximation, see section 11.2.4). This method is, with modifications, in use up till today. Applications for coastal waters, estuaries and lakes followed very soon and apparently independent of one another: Sündermann (1971), Leendertse et al (1973), Heaps (1973), Simons (review 1980). Since that time, the field has developed into an active research area (see e.g. Nihoul & Jamart, 1987, Heaps, 1987). In this chapter, an overview of the numerical treatment of the 3-d SWE is given.

11.2. 3-d Model equations

11.2.1. Momentum and transport

As shown in section 2.4., if you make the shallow-water (or boundary-layer) assumption which says that the vertical length scale is much smaller than the horizontal one, you can conclude that the pressure p must be hydrostatic to a very good approximation:

$$\frac{\partial p}{\partial z} = - \rho g \tag{11.1}$$

or, with the boundary condition $p = p_a$ (atmospheric pressure) at the surface $z = h$:

$$p = p_a + g \int_z^h \rho \, dz \tag{11.2}$$

Here, z is vertical distance, g is the acceleration of gravity and ρ is the density which is now assumed variable. The horizontal momentum equations reduce to (2.16), repeated here for convenience but with some slight modifications:

$$\frac{\partial u}{\partial t} + u \frac{\partial u}{\partial x} + v \frac{\partial u}{\partial y} + w \frac{\partial u}{\partial z} - fv - \frac{\partial}{\partial x}\left(A\frac{\partial u}{\partial x}\right) - \frac{\partial}{\partial y}\left(A\frac{\partial u}{\partial y}\right) - \frac{\partial}{\partial z}\left(v_t\frac{\partial u}{\partial z}\right) =$$

$$= - g \frac{\partial h}{\partial x} - \frac{1}{\rho_0} \frac{\partial p_a}{\partial x} - \frac{g}{\rho_0} \int_z^h \frac{\partial \rho}{\partial x} \, dz \tag{11.3}$$

and similarly for the y-direction. In front of the pressure gradient term, the density has been taken constant, which is the Boussinesq approximation. Turbulent stresses τ_{xz}, τ_{yz} etc. have been expressed in terms of an eddy viscosity v_t (vertically) and A (horizontally), discussed in the next section. The continuity equation is

$$\frac{\partial u}{\partial x} + \frac{\partial v}{\partial y} + \frac{\partial w}{\partial z} = 0 \tag{11.4}$$

The variable density is determined from temperature T, salinity S and possibly pressure, using the equation of state

$$\rho = \rho \, (T, S, p) \tag{11.5}$$

For coastal and estuary flows and other flows considered here, the dependence on pressure can be ignored, which means that water is effectively incompressible. Note that this does not imply that density is constant: it can still vary due to variations in temperature and salinity, which again are determined by a transport equation for heat:

$$\frac{\partial T}{\partial t} + u\frac{\partial T}{\partial x} + v\frac{\partial T}{\partial y} + w\frac{\partial T}{\partial z} - \frac{\partial}{\partial x}\left(D_h\frac{\partial T}{\partial x}\right) - \frac{\partial}{\partial y}\left(D_h\frac{\partial T}{\partial y}\right) - \frac{\partial}{\partial z}\left(D_v\frac{\partial T}{\partial z}\right) = 0 \quad (11.6)$$

and similarly for salt (and possible other substances which might influence density, such as suspended sediment). D_h and D_v are turbulent (eddy) diffusion coefficients.

11.2.2. Parameterizations

Even though the full 3-d flow is now represented, you still need parameterizations for the turbulent transport terms. In (11.3) and (11.6), a parameterization in terms of eddy diffusion has been assumed. It can be argued that eddy transports in horizontal and vertical directions are of similar magnitude, with the consequence that horizontal transport terms could be neglected due to the much larger length scales involved. Yet, they are often retained for numerical stability and smoothness. Moreover, the discretized equations to be discussed later can be considered as averages over grid intervals. The stress terms then represent *subgrid-scale* transport which may be relevant in at least some applications (e.g. coarse-grid models in oceanography). These terms are represented by horizontal viscosity and diffusion terms with coefficients A and D_h. The value of these is very uncertain, but often not very important.

Smagorinsky (1963) proposed to express subgrid-scale transports by an effective eddy-viscosity related to the grid size. The eddy viscosity is assumed to be determined by the grid size and the resolved velocity gradients. The idea is related to the mixing-length approach for vertical eddy transport (11.9) discussed below. On dimensional grounds, the horizontal eddy viscosity should then look like

$$A = c_A \Delta x^2 \sqrt{S_{ij}S_{ij}}$$

where repeated subscripts imply summation, c_A is a constant and the deformation rate is given by

$$S_{ij} = \frac{\partial u_i}{\partial x_j} + \frac{\partial u_j}{\partial x_i} \quad (i, j = 1,2)$$

This leads to a relatively complicated expression. In many applications, just a constant, empirically estimated, value of A is used.

Much more can be said about the *vertical turbulent transport* terms. This is where you enter turbulence theory and only a few results are presented here. For more details and background you should consult Rubesin (1977), Rodi (1980) or Launder (1989). These turbulent transports are essential for the flow structure. A hierarchy of models is available to model the eddy viscosity v_t and similarly D_v.

a. *Constant eddy viscosity*, that is independent of vertical position z. You could call this the quasi-laminar approximation. The magnitude of eddy viscosity is related to the typical shear stress

$$v_t = c_0 u_* a \sim c_0 c_f V a \quad (11.7)$$

where $u*$ is the shear velocity, V the typical mean velocity and a the depth. This could result in values of the order 0.05 m²/s. Using such an expression means that the viscosity does vary with x,y,t. This approach may work well in some cases if the magnitude is well chosen (Davies & Gerritsen, 1994), but velocity profiles will generally not look very realistic. For example, in steady, uniform flow, a parabolic velocity profile will be obtained.

b. *Specified vertical profile* of eddy viscosity. It turns out that (in homogeneous flow) a linear variation near the bottom is important. As a generalization of (11.7), you could take

$$v_t = 0.2 \; \kappa u*a \qquad (z - z_b > 0.2 \; a)$$
$$= \kappa u*(z - z_b + z_0) \qquad (z - z_b < 0.2 \; a) \tag{11.8}$$

where z_0 is a parameter related to the bottom roughness, z_b is the bottom level and κ (= 0.4) is Von Karman's constant . This type of formulation produces logarithmic velocity profiles near the bottom, in qualitative agreement with observations. The approach will not be satisfactory for stratified flows, however. Neither does it include any influence from wind at the surface.

c. *Mixing length.* The eddy viscosity is supposed to depend on the local velocity gradient and a mixing length l:

$$v_t = l^2 \sqrt{u_z^2 + v_z^2} \; f(Ri) \tag{11.9}$$

The mixing length could be related to the distance to the nearest wall, e.g. $l = \kappa (z - z_b + z_0)$. The Richardson number, defined as

$$Ri = - \frac{\dfrac{g}{\rho} \dfrac{\partial \rho}{\partial z}}{u_z^2 + v_z^2}$$

measures the influence of density stratification. If it exceeds a critical value (say, 0.5), turbulence is supposed to be fully suppressed. The function $f(Ri)$ therefore is usually chosen as a negative exponential leading to this suppression. For applications, see Perrels & Karelse (1981) and for homogeneous flow Davies and Gerritsen (1994). Note that the turbulent-stress terms in the momentum equations now become quadratic in the velocity gradients.

d. *Turbulent energy model.* Turbulence should be parameterized by using at least two quantities: a length scale (such as the mixing length above) and an intensity. On dimensional grounds, the eddy viscosity should then look like

$$v_t = c_v k^{1/2} l \tag{11.10}$$

with some constant c_v. It is possible to derive an equation for the intensity or kinetic energy k of turbulent fluctuations (Rodi, 1980), in boundary-layer form:

$$\frac{\partial k}{\partial t} + u \frac{\partial k}{\partial x} + v \frac{\partial k}{\partial y} + w \frac{\partial k}{\partial z} - \frac{\partial}{\partial z}\left(D_k \frac{\partial k}{\partial z}\right) = \frac{\tau_{xz}}{\rho} \frac{\partial u}{\partial z} + \frac{\tau_{yz}}{\rho} \frac{\partial v}{\partial z} + g \, K_v \frac{\partial \rho}{\partial z} - \varepsilon \quad (11.11)$$

where D_k is another turbulent diffusion coefficient of the same order as v_t. The terms in (11.11) can be interpreted as advection of energy, turbulent diffusion and, on the right-hand side, production of turbulent energy by work of the shear stresses τ against velocity gradients, conversion of kinetic into potential energy through diffusion of heat and salinity (i.e. density) and ε is the rate of energy conversion into heat. The latter can be argued to be dependent on the local turbulence intensity and the length scale as

$$\varepsilon = c_\varepsilon \frac{k^{3/2}}{l} \tag{11.12}$$

The idea is to solve (11.11) together with the model equations mentioned in section 11.2.1. The length scale could be specified as a mixing length. As one differential equation is used to characterize turbulence, this is sometimes called a one-equation turbulence model.

e. *k-ε model.* An extension of the previous model is to use (11.12) as a *definition* of the length scale and to solve a differential equation for ε as well. This gives a two-equation turbulence model. Note that the numerical effort increases significantly by adding these equations which are at least as complicated as the original momentum equations. The ε equation is much less certain than the energy equation but it is generally taken to be of the following form

$$\frac{\partial \varepsilon}{\partial t} + u \frac{\partial \varepsilon}{\partial x} + v \frac{\partial \varepsilon}{\partial y} + w \frac{\partial \varepsilon}{\partial z} - \frac{\partial}{\partial z}\left(D_\varepsilon \frac{\partial \varepsilon}{\partial z}\right) = \frac{c_1 \varepsilon}{k}\left(\frac{\tau_{xz}}{\rho}\frac{\partial u}{\partial z} + \frac{\tau_{yz}}{\rho}\frac{\partial v}{\partial z} + g\, D_v \frac{\partial \rho}{\partial z}\right) - c_{2\varepsilon}\frac{\varepsilon^2}{k} \tag{11.13}$$

which includes a number of additional constants. Other definitions of length scales are also possible; see e.g. Mellor & Yamada (1982), Sheng (1987).

Even more complicated models of turbulence exist but have not been shown to give a clear benefit in return for the increased effort. The k-ε model is about the most elaborate model that has been used so far and even so there is no common opinion on the relative merits of the various models. Several authors suggest that for simple situations (no density stratification involved) models of the categories b/c might be sufficiently accurate.

11.2.3. Boundary conditions

Surface- and bottom boundary conditions have already been discussed in section 2.2 but are repeated here for convenience and specialized to the boundary-layer approximation and the eddy-viscosity parameterization. At the free surface $z = h(x,y,t)$ you have three conditions:

kinematic (2.9) $\qquad \dfrac{\partial h}{\partial t} + u \dfrac{\partial h}{\partial x} + v \dfrac{\partial h}{\partial y} - w = 0$ $\qquad\qquad$ (11.14)

dynamic: pressure $\qquad p = p_a$ $\qquad\qquad\qquad\qquad\qquad\qquad\qquad\qquad$ (11.15) dynamic: shear stress (2.12) $\dfrac{(\tau_{sx}, \tau_{sy})}{\rho} = v_t \dfrac{\partial}{\partial z}(u,v)$ (11.16)

At the bottom $z = z_b(x,y)$, there are two conditions:

kinematic (2.8) $u\dfrac{\partial z_b}{\partial x} + v\dfrac{\partial z_b}{\partial y} - w = 0$ (11.17)

dynamic: no-slip $u = v = 0$ (11.18)

The total differential order of the momentum, pressure and continuity equations in z-direction is 4, so you would expect to have 4 boundary conditions. That you have 5 is because the *level* of the free surface h is still unknown. The additional boundary condition serves to fix it either explicitly or implicitly.

Formally, anyone will agree that (11.18) is the correct dynamic condition at the bottom. It may be somewhat awkward to handle numerically, however, because it usually leads to very strong velocity gradients (logarithmic profile) near the bottom, which are difficult to resolve and often not really interesting. The quantity you are interested in is usually the shear stress at the bottom, e.g. to compute sediment transport. A common alternative in turbulence modelling is using the *law of the wall* (also indicated as drag law, partial-slip or mixed boundary condition). It assumes that very near the bottom a constant-stress layer exists in which a logarithmic velocity profile can be deduced using the mixing-length model discussed above. The velocity at the margin of that layer, that is at some distance δ from the bottom, can then be related to the bottom stress, which by assumption equals the stress at level δ:

$$\dfrac{(\tau_{zx}, \tau_{zy})}{\rho} = v_t \dfrac{\partial}{\partial z}(u,v) = c_f(u,v) \quad \text{at} \quad z = z_b + \delta \tag{11.19}$$

where c_f is a drag coefficient depending on the bottom roughness. Effectively, the boundary of the numerical model is now at $z_b + \delta$ and the boundary condition (11.19) is of mixed type: it contains both (u,v) and its vertical derivative.

The boundary conditions for the 3-d SWE at side walls are analogous to those for the 2-d equations (chapter 5) but the theory is much less complete (see Oliger & Sundström,1978).

11.2.4. Surface level

In principle, the surface level h can be computed from (11.14), once the velocity field is known from the momentum and continuity equations. However, it is not very common to do so. A more robust equation is obtained by integrating the continuity equation (11.4) over depth, which gives (2.17):

$$\dfrac{\partial h}{\partial t} + \dfrac{\partial}{\partial x}(a\,\overline{u}) + \dfrac{\partial}{\partial y}(a\,\overline{v}) = 0 \tag{11.20}$$

This again determines the rate of change of h if the velocities are known. The latter, of course, are again influenced by the surface level. To take this feedback into account, you can also integrate the momentum equations over depth, which gives

$$\frac{\partial}{\partial t}(a\,\overline{u}) + ga\,\frac{\partial h}{\partial x} = R_x$$

$$\frac{\partial}{\partial t}(a\,\overline{v}) + ga\,\frac{\partial h}{\partial y} = R_y$$

(11.21)

where the right-hand sides contain all integrated terms not explicitly shown. Combining this with (11.20) gives the wave equation (4.6) generalized to three dimensions. Supposing the right-hand sides in (11.21) to be known (e.g. from the preceding time step), you can solve (11.20) together with (11.21) in every time step for the surface level h, which takes the main feedback with velocity into account. Note that this is a numerical process in only two space dimensions. Consequently, it will take relatively little effort compared with the 3-d momentum equations. It is possible to include the Coriolis terms in this depth-averaged procedure, if they are important (Blumberg & Herring, 1987). For steady-state flows, the procedure reduces to a Poisson equation for h.

Some simplifications are still possible. Firstly, if the water depth does not change very much, you could use the mean or initial depth on the left-hand side of (11.20) and (11.21). The error between it and the actual depth could be corrected on the right-hand side. This simplifies the numerical solution considerably because the matrix becomes constant; only the right-hand side changes per time step.

In some cases, the water-level changes may be so small that you are not interested in them. You can then introduce the *rigid-lid approximation* which assumes the volumetric effect of $\partial h/\partial t$ to be small compared with the other terms in the integrated continuity equation (11.20). The free surface is replaced by a rigid lid. The time derivatives in (11.14) or (11.20) then disappear. You may *not*, however, neglect the surface slopes $\partial h/\partial x$, $\partial h/\partial y$ in the momentum equations. A rigid lid will support a nonzero pressure which can be related to a virtual water level h. You therefore still have to compute h from (11.20) (without time derivative) and (11.21) and the result represents the pressure exerted on the rigid lid. This approximation is not uncommon in oceanography, where the free-surface changes are extremely small compared with actual depth (e.g. Bryan, 1969 who avoids computing the surface level but rather determines a stream function for the depth-averaged flow).

11.2.5. σ - coordinates

The domain in which the 3-d SWE are to be solved varies not only in space (by bottom topography) but also in time (by free-surface movements). This introduces numerical difficulties. If you use a fixed numerical grid, the thickness of the upper layer varies. This can be handled if the water-level variation does not exceed the surface-layer thickness; otherwise the top layer will disappear and the second layer gets a variable thickness. Conversely, this means that the vertical grid size would have to be chosen in excess of the expected water-level variation which will be unacceptable in many cases. In such cases a very useful device is the use of the σ-transformation. It is generally ascribed to Phillips (1957) and is related to similar transformations used in meteorology, where pressure is used as a vertical coordinate. The idea is simply to normalize the vertical dimension to unity by the transformation

$$\sigma = \frac{z - z_b(x,y)}{a(x,y,t)} \qquad\qquad x' = x \tag{11.22}$$

where σ varies between 0 and 1 irrespective of free-surface movements. The computational domain is then fixed; see fig. 11.1. You could also say that the grid points move up and down with the surface level. The price to be paid is that the differential equations get slightly more complicated. Also, it is obvious that difficulties arise if the water depth goes to zero anywhere. The coordinate transformation implies relations like

$$\frac{\partial u}{\partial z} = \frac{1}{a}\frac{\partial u}{\partial \sigma}$$

$$\frac{\partial u}{\partial x} = \frac{\partial u}{\partial x'} - \frac{1}{a}\left(\frac{\partial z_b}{\partial x} + \sigma\frac{\partial a}{\partial x}\right)\frac{\partial u}{\partial \sigma}$$

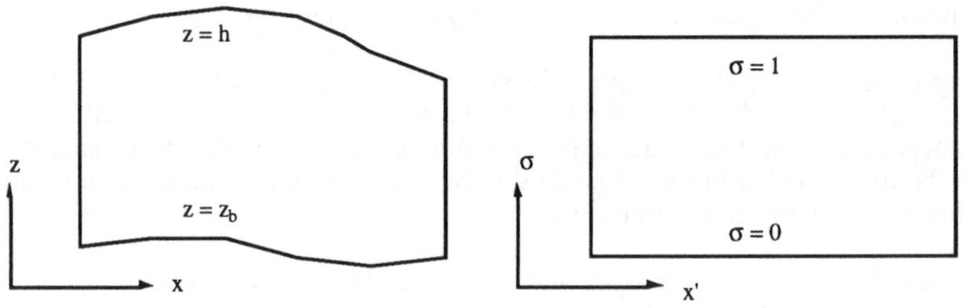

Fig. 11.1. Physical domain and σ -transformation

Substituting these into the differential equations (11.1) and (11.3) gives (omitting the prime on x')

$$\frac{1}{a}\frac{\partial p}{\partial \sigma} = -\rho g \tag{11.23}$$

$$\frac{\partial u}{\partial t} + u\frac{\partial u}{\partial x} + v\frac{\partial u}{\partial y} + \omega\frac{\partial u}{\partial \sigma} - fv \;\; - \frac{1}{a}\frac{\partial}{\partial \sigma}\left(\frac{v_t}{a}\frac{\partial u}{\partial \sigma}\right) = -\frac{1}{\rho}\frac{\partial p}{\partial x} + \frac{1}{\rho a}\left(\frac{\partial z_b}{\partial x} + \sigma\frac{\partial a}{\partial x}\right)\frac{\partial p}{\partial \sigma} \tag{11.24}$$

where horizontal viscosity terms have been omitted for simplicity. They could be added again; however, considering their *ad-hoc* character it does not seem to make sense to apply the formal coordinate transformation to them. Actually, Mellor & Blumberg (1985) advocate replacing the vertical diffusion term in (11.3) by one in σ direction and leaving the horizontal components untransformed. Otherwise, in case of relatively strong bottom slopes, an unrealistic component of the "vertical" turbulent transport in (11.3) parallel to the bottom would result. Note that a modified "vertical" velocity has been defined which is actually the velocity across a level of constant σ.

$$\omega = \frac{1}{a}\left[w - u\frac{\partial z_b}{\partial x} - v\frac{\partial z_b}{\partial y} - \sigma(\frac{\partial a}{\partial t} + u\frac{\partial a}{\partial x} + v\frac{\partial a}{\partial y})\right] \qquad (11.25)$$

Using this, (11.24) is very similar to the original equation (11.3). Transforming the continuity equation (11.4) gives

$$\frac{\partial u}{\partial x} + \frac{\partial v}{\partial y} + \frac{1}{a}\left[\frac{\partial w}{\partial \sigma} - (\frac{\partial z_b}{\partial x} + \sigma\frac{\partial a}{\partial x})\frac{\partial u}{\partial \sigma} - (\frac{\partial z_b}{\partial y} + \sigma\frac{\partial a}{\partial y})\frac{\partial v}{\partial \sigma}\right] = 0$$

Keeping in mind that z_b and a do not depend on σ and using the definition of ω, this can be rewritten as

$$\frac{\partial u}{\partial x} + \frac{\partial v}{\partial y} + \frac{\partial \omega}{\partial \sigma} + \frac{1}{a}\left[\frac{\partial a}{\partial t} + u\frac{\partial a}{\partial x} + v\frac{\partial a}{\partial y}\right] = 0$$

or, after multiplication by a

$$\frac{\partial a}{\partial t} + \frac{\partial}{\partial x}(au) + \frac{\partial}{\partial y}(av) + \frac{\partial}{\partial \sigma}(a\omega) = 0 \qquad (11.26)$$

which replaces (11.4). Note that this is a local, not an integrated continuity equation, even though it looks superficially like (11.20). You can conclude that the equations in σ-coordinates are only slightly more complicated than those in Cartesian coordinates. One of the advantages is that the kinematic boundary conditions at surface and bottom become extremely simple:

$$\text{at } \sigma = 0: \quad \omega = \frac{1}{a}\left[w - u\frac{\partial z_b}{\partial x} - v\frac{\partial z_b}{\partial y}\right] = 0 \qquad (11.27)$$

$$\text{at } \sigma = 1: \quad \omega = \frac{1}{a}\left[w - u\frac{\partial z_b}{\partial x} - v\frac{\partial z_b}{\partial y} - (\frac{\partial a}{\partial t} + u\frac{\partial a}{\partial x} + v\frac{\partial a}{\partial y})\right] =$$
$$= \frac{1}{a}\left[w - (\frac{\partial h}{\partial t} + u\frac{\partial h}{\partial x} + v\frac{\partial h}{\partial y})\right] = 0 \qquad (11.28)$$

The surface level h can now only be determined from the *depth-integrated* continuity equation (11.26), which, not surprisingly, is identical to (11.20).

11.2.6. Why not the full 3-d Navier-Stokes equations?

Of course, you can solve all 3-d problems considered here by the full 3-d Navier-Stokes equations. The great advantage is that you do not have to worry about any approximations concerning horizontal and vertical length scales and neglect of vertical accelerations. You can take any complicated geometry and any hydraulic structure into account, including such things as flow separation. Such methods do exist (see, e.g. Peyret & Taylor, 1983, Fletcher, 1988, Versteegh, 1990).

However, solving a 3-d Navier-Stokes problem is computationally more demanding than solving the 3-d SWE for at least two reasons (supposing of course that conditions for the validity of the SWE are met; otherwise you do not have a choice). Firstly, in the former case, you will have to solve a 3-d Poisson equation for the pressure distribution in each time step; in the SWE case this is replaced by a 2-d problem (section 11.2.4). Secondly, it is not at all clear that you can use elongated grid cells (high aspect ratio $\Delta x/\Delta z$) in a Navier-Stokes code for accuracy reasons. You might be forced to use considerably more grid points than in the SWE case. Therefore, for large-(horizontal-) scale problems with a 3-d structure, 3-d SWE methods will continue to be attractive.

11.3. Discretization in space

It is general practice to discretize horizontal and vertical processes in the SWE separately, possibly by different grids or methods. Even if grid transformations (curvilinear or boundary-fitted grids) are applied, this is usually done separately.

11.3.1. Horizontal discretization

In the horizontal x-y plane, you have the same possibilities for discretization as in the case of the 2-d SWE (chapter 6). The properties concerning stability and accuracy will also be generally the same.

Among the *finite-difference* grids, the C-grid is by far the most popular due to its relative efficiency and lack of sensitivity to grid-size wiggles. For regions of complicated geometry, curvilinear or boundary-fitted grids are sometimes used (e.g. Blumberg & Herring, 1987, Sheng, 1987).

Finite-element methods are not very common (as in the 2-d case), probably due to their tendency to be relatively expensive.

Spectral methods suffer from the same problem as in the 2-d case, that they are difficult to apply on non-rectangular regions. In global meteorology, with periodic boundary conditions, spectral methods are very suitable and therefore used in most numerical weather-prediction models. For closed basins, as in (coastal) oceanography, a transformation to a rectangular region by curvilinear coordinates is needed before spectral methods can be used. This has been done by Haidvogel et al (1991). It yields a very powerful method in principle, although the coordinate transformation can be a problem for really complicated geometries.

11.3.2. Vertical discretization

The new aspect of 3-d SWE is, of course, the vertical dimension. Again, you have similar choices as for the horizontal discretization. However, you need not necessarily make the same choice. For example, it is not uncommon to use spectral methods horizontally and finite-difference or finite-element methods vertically (particularly in meteorological models). Conversely, a regular horizontal finite-difference method may be combined with a spectral approach vertically (see below).

Quite a number of methods work on a full 3-d C-grid (fig. 11.2). If applied in physical coordinates, the bottom will usually not be a grid line. Consequently, you must approximate it by a "staircase" type boundary, like fig. 6.4. This is not a very accurate approximation and numerical noise may result near the "steps". On the other hand, it is efficient as shallow areas are reproduced with fewer points than deeper ones, which seems useful from a physical point of view. A similar problem is encountered at the surface. If water-level variations are very small, you might apply the surface boundary conditions in a linearized form, i.e. at level $z = 0$ (undisturbed water level). This is often done in oceanographic applications. However, if depth variations are important, which is normally the case in coastal areas or estuaries, you must allow the top layer to be of variable depth and thicker than the maximal variation of water level (Leendertse et al, 1973, Leendertse, 1989).

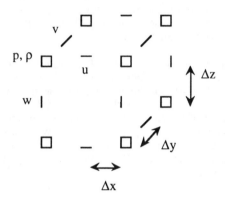

Fig. 11.2. Three-dimensional C-grid

In many cases, particularly in oceanographic applications, it is customary to use a vertically non-uniform grid: surface layers are much thinner than deeper layers. This may have drawbacks for numerical accuracy. For example, on an irregularly spaced grid it is possible to approximate first derivatives with second-order accuracy, but not second derivatives:

$$\frac{\Delta z_{j-1}}{\Delta z_{j-1} + \Delta z_j} \frac{u_{j+1} - u_j}{\Delta z_j} + \frac{\Delta z_j}{\Delta z_{j-1} + \Delta z_j} \frac{u_j - u_{j-1}}{\Delta z_{j-1}} = \frac{\partial u}{\partial z} + O(\Delta z^2)$$

$$\frac{\Delta z_{j-1}(u_{j+1} - u_j) - \Delta z_j(u_j - u_{j-1})}{2\Delta z_j \Delta z_{j-1}(\Delta z_{j-1} + \Delta z_j)} = \frac{\partial^2 u}{\partial z^2} + O(\Delta z_j - \Delta z_{j-1})$$

where $\Delta z_j = z_{j+1} - z_j$. Therefore, if viscosity is important, you will obtain only first-order accuracy on an arbitrarily spaced grid. However, if viscosity is unimportant or poorly known, this drawback might not be so serious. Moreover, if the grid is non-uniform but varying in a regular way, that is $\Delta z_j = \Delta z_{j-1} (1 + \kappa \Delta z_{j-1})$ with some constant κ, you do get second-order accuracy even for the second derivatives (verified by Davies, 1991).

Similar discretizations can also be applied in σ-coordinates. The bottom and surface are then grid lines and therefore represented almost ideally (if the horizontal grid is sufficiently fine, of course). In this case, again, vertically non-uniform grids may be used (Davies, 1991). One way to define these more formally is to replace the σ-coordinate in (11.22) by

$$\sigma = \phi \left(\frac{z - z_b(x,y)}{a(x,y,t)} \right)$$

with some suitably chosen, smooth, function ϕ which expands or compresses certain regions, wherever needed.

Finite elements are sometimes but not very often used. They may not be expected to be essentially more accurate than finite-difference methods, but they offer a good possibility of using vertically non-uniform grids in a systematic way. This seems to be the major advantage. For an example see Robert & Ouellet (1987).

Standard spectral methods (using, e.g. trigonometric of Chebyshev functions as basis functions) are normally considered too complicated as they lead to numerical coupling between all components. This is a drawback compared with finite-difference or finite-element methods, where usually each layer is coupled only to its direct upper and lower neighbours. However, a special type of spectral methods has been proposed by Heaps (1973) and elaborated by Davies in a series of papers (good review in Davies, 1987), in which this coupling problem is avoided. The method has, however, some limitations. First of all, it appears to have been used for homogeous flows only. Secondly, it assumes that the vertical viscosity can be separated as

$$\nu_t(x,y,z,t) = N(x,y,t)\ Z(\sigma)$$

(note σ, not z in the second component) which implies that the vertical profile is similar at all horizontal positions. Some of the simpler turbulence models discussed in section 11.2.2 can be written in this form. The third limitation is that advection terms are considered unimportant. If they are neglected, the vertical velocity is not needed if you use the depth-averaged continuity equation (11.20). The velocity components are now approximated as

$$u = \sum_{j=1}^{n} U_j(x,y,t)\ \phi_j(\sigma)$$

$$v = \sum_{j=1}^{n} V_j(x,y,t)\ \phi_j(\sigma)$$

where the functions ϕ are defined to be eigenfunctions of the following equation with corresponding eigenvalue λ:

$$\frac{d}{d\sigma} \left(Z \frac{d\phi_j}{d\sigma} \right) = - \lambda_j \phi_j \tag{11.29}$$

This equation can be solved separately, using some suitable method. The eigenfunctions are orthogonal if the eigenvalues are different, which is shown as follows.

$$\int_0^1 \phi_k \phi_j \, d\sigma = -\frac{1}{\lambda_k} \int_0^1 \phi_j \frac{d}{d\sigma} \left(Z \frac{d\phi_k}{d\sigma} \right) d\sigma = \frac{1}{\lambda_k} \int_0^1 Z \frac{d\phi_k}{d\sigma} \frac{d\phi_j}{d\sigma} \, d\sigma$$

by partial integration, where the following boundary conditions are assumed for all k:

$$\sigma = 0 \quad \phi_k = 0$$

$$\sigma = 1 \quad \frac{d\phi_k}{dz} = 0$$

(11.30)

However, by the same reasoning, you derive

$$\int_0^1 \phi_k \phi_j \, d\sigma = \frac{1}{\lambda_j} \int_0^1 Z \frac{d\phi_k}{d\sigma} \frac{d\phi_j}{d\sigma} \, d\sigma$$

so the integral must be zero if the eigenvalues are different. The eigenfunctions can be normalized in some way, e.g.

$$\int_0^1 \phi_k^2 \, d\sigma = 1$$

(this is different from the normalization used by Davies). The Galerkin approach is now used to generate (vertically) discretized equations: the momentum equations (11.24) are multiplied by ϕ_k and integrated from bottom to top. For horizontal discretization you can then subsequently use any of the methods discussed above. In continuous form, this gives

$$\frac{\partial U_k}{\partial t} - fV_k + \frac{N \lambda_k}{a^2} U_k = g \, \alpha_k \frac{\partial h}{\partial x} \qquad k = 1,...,n$$

(11.31)

(similarly for v) where

$$\alpha_k = \int_0^1 \phi_k \, d\sigma$$

The integrated continuity equation (11.20) becomes

$$\frac{\partial h}{\partial t} + \sum_{k=1}^n \left\{ \frac{\partial}{\partial x} (\alpha_k a U_k) + \frac{\partial}{\partial y} (\alpha_k a V_k) \right\} = 0$$

(11.32)

The component-momentum equation (11.31) is fully uncoupled from the other components; it comprises a linear, Newton- or Rayleigh-type frictional term which can be easily accounted for numerically. Coupling between the components is only by means of the water level h and the continuity equation (11.32).

An example of the basisfunctions is given in fig. 11.3 for the test problem discussed in section 11.6.1. Davies claims that a good accuracy is already obtained using as few as 5 components, as compared to some 20 grid points in vertical direction for finite-difference methods. This agrees with the general behaviour of spectral methods.

The boundary conditions (11.30) are consistent with no-slip on the bottom and free-slip at the surface. However, if you would want to apply different boundary conditions, the basis functions must be subject to different boundary conditions as well. For example, if wind stress at the surface is to be taken into account, neither ϕ nor its vertical derivative may be taken zero at the surface. This means, however, that the basis functions will no longer be orthogonal and all component-momentum equations will be coupled (Davies, 1987). Similar complications occur for more general viscosity formulations, taking advection terms into account and coupling with density variations. Therefore, this spectral method appears to be very effective but of somewhat limited application.

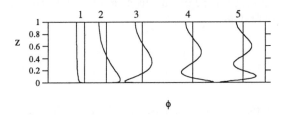

Fig. 11.3. Spectral functions for linear viscosity profile (see section 11.6.1)

11.4. Discretization in time

Due to the relative complexity of the 3-d SWE and boundary conditions (including a moving boundary), one has generally tried to apply as simple discretization methods as possible. A typical example is a fully explicit leap-frog (LF) method on a 3-d C-grid, which was used, e.g. by Leendertse (1973). As the equations contain both advection and diffusion terms, a straightforward LF method will not work as it is unstable with diffusion included. However, if the diffusion terms are evaluated either at the previous time level $n-1$ or at the new time level $n+1$, a conditionally stable method results. You get a method very similar to the 2-d version (6.9). For homogeneous flow:

$$D_{0t}u + u\,M_xD_{0x}u + (M_xM_yv\,)M_yD_{0y}u + (M_xM_zw)\,M_zD_{0z}u +$$
$$- f\,M_xM_yv - D_{0x}(A\,D_{0x}u\,) - D_{0y}(A\,D_{0y}u\,) - D_{0z}(v_t\,D_{0z}u\,) = g\,D_{0x}h \qquad (11.33)$$

where various averages are needed due to the staggered grid. Diffusion terms should be evaluated at n-1; all other terms at n. The continuity equation becomes straightforwardly

$$D_{0x}u + D_{0y}v + D_{0z}w = 0 \qquad (11.34)$$

which is used to compute w, once u and v are known from the momentum equations. The water level h is computed from (11.20) in LF approximation.

A complete stability analysis of this system including boundary conditions and non-linearities is virtually impossible. From simplified cases, you can guess the type of stability conditions to be expected. For example, in one space dimension, the LF method for an advection-diffusion equation with constant coefficients will be stable if

$$0 < \lambda < \frac{1}{2}(1 - \sigma^2)$$

where $\lambda = 2D \, \Delta t / \Delta x^2$ and $\sigma = u \, \Delta t/\Delta x$ is the Courant number. Taking into account information from the 2-d SWE, the following restrictions are expected (Table 11.1).

Table 11.1. Stability restrictions for explicit method

Physical process	Stability condition	Typical Δt (s)
external (surface) wave	$\Delta t < \dfrac{\Delta x}{\sqrt{2ga}}$	400
internal wave	$\Delta t < \dfrac{\Delta x}{\sqrt{2\varepsilon ga}}$	6000
Coriolis (inertial wave)	$\Delta t < 1/f$	10^4
horizontal diffusion	$\Delta t < \dfrac{\Delta x^2}{4A}$	$2.5 \ 10^6$
vertical diffusion	$\Delta t < \dfrac{\Delta z^2}{4v_t}$	100
horizontal advection	$\Delta t < \dfrac{\Delta x}{U}$	10^4
vertical advection	$\Delta t < \dfrac{\Delta z}{w}$	$4 \ 10^4$

Here, ε is a typical relative density difference $\Delta\rho/\rho$. The values in the last column have been computed for the following example situation (shallow sea):

$$
\begin{array}{ll}
a & = 30 \text{ m} \\
\varepsilon & = 0.005 \\
f & = 10^{-4} \text{ s}^{-1} \\
A & = 10 \text{ m}^2/\text{s} \\
v_t & = 0.01 \text{ m}^2/\text{s} \\
u & = 1 \text{ m/s} \\
w & = 5 \ 10^{-5} \text{ m/s} \\
\Delta x & = 10 \text{ km} \\
\Delta z & = 2 \text{ m}
\end{array}
$$

Even though this is only an example, it is obvious that there are two strong restrictions: surface waves and vertical viscosity (diffusion). The other limitations are not really restrictive as the time step could not be very much larger than 0.5 h anyway, due to external conditions (tidal period, variation of wind stress etc.).

A fully implicit method would be hardly feasible. Therefore, *semi-implicit* methods have been constructed in which the two limiting factors have been eliminated. Firstly, vertical diffusion can be handled implicitly (level $n+1$ in 11.33, see, e.g. Blumberg & Mellor, 1987). Implicit coupling is then obtained between u-values in each column of grid points separately (similarly for v). Together with the surface and bottom boundary conditions, this results in a simple tridiagonal system of equations per column, which can be solved efficiently using the Thomas or double-sweep algorithm. On a vector computer, one would vectorize over the horizontal grid (that is, do all the tridiagonal systems in parallel).

Secondly, the computation of the free surface can be done implicitly. To this end, you use finite-difference forms of (11.20) and (11.21) where the spatial derivatives on the left-hand side are averaged between levels $n+1$ and $n-1$. You obtain a 2-d system of equations for water level and depth-averaged velocities, which is very similar to (8.22) (Elvius & Sundström) and can be solved either directly or iteratively. Once the water-level has been determined, you can use (11.33) to determine the velocity distribution (viscosity either explicit or implicit). If the discretization has been done consistently, the integral (discrete sum) of the velocities will agree with the already computed depth-average values.

Some other methods have been proposed. Leendertse (1989) developed an *ADI* method in which horizontal directions are treated implicitly in alternating order. Advection is treated by higher-order upstream methods proposed by Stelling (1983), see section 11.5. The method is claimed to be at least second-order accurate, contrary to some of the above-mentioned methods which include first-order accurate terms. An outline of the method follows.

Each time step consists of two stages. In the first stage, some of the x-derivatives are treated implicitly. The v-momentum equation with implicit vertical diffusion is (with A and D_h indicating advection and horizontal diffusion terms)

$$
\frac{v^{n+1/2} - v^n}{\Delta t/2} + A(v^n) - D_h(v^n) - D_{0z}(v_t \, D_{0z}v^{n+1/2}) = g \, D_{0y}h^n \tag{11.35}
$$

This consists of a number of tridiagonal system per vertical column, which can be solved in parallel if the fields from the previous time step are known. The u-momentum equation is treated similarly but with implicit pressure gradient:

$$\frac{u^{n+1/2} - h^n}{\Delta t/2} + A(u^*) - D_h(u^*) - D_{0z}(v_t D_{0z}u^{n+1/2}) = g D_{0x}h^{n+1/2} \qquad (11.36)$$

The terms with an asterisk are discretized either on old or new time levels, depending on the order in which the grid points are met; for details see the original publications. If you add (11.36) over the grid points in a vertical column, you obtain an equation in terms of $h^{n+1/2}$, $\bar{u}^{n+1/2}$, which can be combined with the depth-integrated continuity equation

$$\frac{h^{n+1/2} - h^n}{\Delta t/2} + D_{0x}(M_x a^n \bar{u}^{n+1/2}) + D_{0y}(M_y a^n \bar{v}^n) = 0 \qquad (11.37)$$

You obtain systems of tridiagonal equations along rows of grid points which can be solved for $h^{n+1/2}$, $\bar{u}^{n+1/2}$. Using these results, (11.36) can be solved per column (again tridiagonal systems) for the individual velocities u. This completes the first stage. In the second stage, the role of x- and y-directions is interchanged (compare (8.27)). For the linear terms, second-order accuracy in time is obtained over a full time step. Using a special choice for the advective and horizontal diffusion terms, Leendertse obtains second-order accuracy for those as well.

Usseglio-Polatera & Sauvaget (1987) extended the *fractional-step* idea of section (8.6) to three dimensions. No less than nine fractional steps are introduced, even without density gradients being taken into account:
1. advection of depth-averaged momentum,
2. horizontal advection of local momentum,
3. vertical advection of local momentum (all advection steps using a characteristics-based method),
4. diffusion of depth-averaged momentum (including Coriolis terms), using an ADI method,
5. horizontal diffusion of local momentum,
6. vertical diffusion of local momentum,
7. determination of water-level, using (11.20), (11.21) solved iteratively by a conjugate-gradient method,
8. specification of velocity profiles on open boundaries and interfaces to surrounding depth-averaged (2-d) model,
9. update of local velocities to pressure gradient.

Due to this splitting procedure, which is not reversed in the next time step, the method must involve a first-order time-discretization error. The method is too complicated to be presented here.

11.5. Advection-diffusion

In case of stratified flows, density is determined by salinity S and temperature T, both of which are transported with the flow. You therefore need to solve transport equations of the form

$$\frac{\partial c}{\partial t} + \frac{\partial}{\partial x}\left(uc - D_h\frac{\partial c}{\partial x}\right) + \frac{\partial}{\partial y}\left(vc - D_h\frac{\partial c}{\partial y}\right) + \frac{\partial}{\partial z}\left(wc - D_v\frac{\partial c}{\partial z}\right) = 0 \tag{11.38}$$

where c may be temperature or salinity or the concentration of any other substance carried by the flow (pollutants, suspended sediment, etc.). Actually, advection of momentum can be treated the same way. D_h and D_v are (eddy) diffusion coefficients in horizontal and vertical directions. They are often given different values due to differences in resolved and unresolved scales. This type of advection-diffusion equations has been the subject of a tremendous number of studies. It turns ot be difficult to construct a method that satisfies a number of criteria simultaneously: besides stability and accuracy it is usually required that
1. Total mass (or heat) be conserved in some sense,
2. No negative concentrations occur,
3. No numerical oscillations or wiggles occur.
Consequently, a great number of numerical methods exist that try to combine these properties. Reviews are given, a.o. by Chock & Dunker (1983 - 1991), Rood (1987) and Vreugdenhil & Koren (1993). Here only a brief overview can be given. You will note that there is a correspondence with Riemann solvers (section 8.7).

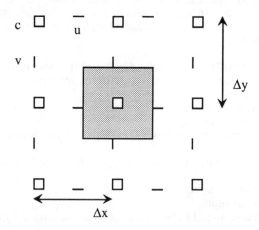

Fig. 11.4. Definition of control volume

A useful framework is the finite-volume approach, in which the conservation equation (11.38) is integrated over a control volume of size Δx, Δy, Δz surrounding a particular grid point i, j, k: (fig. 11.4)

$$\frac{\partial c}{\partial t} + \frac{F_{i+1/2,j,k} - F_{i-1/2,j,k}}{\Delta x} + \frac{G_{i,j+1/2,k} - G_{i,j-1/2,k}}{\Delta y} + \frac{H_{i,j,k+1/2} - H_{i,j,k-1/2}}{\Delta z} = 0 \tag{11.39}$$

where F, G, H are the fluxes shown in (11.38), e.g.

$$F = uc - D_h \frac{\partial c}{\partial x}$$

The control-volume form of (11.39) ensures that, whatever approximation you use for the fluxes, total conservation will be satisfied because the flux leaving one cell is exactly the same as the flux entering the neighbour cell. The fluxes are defined halfway the grid points and must now be expressed in terms of grid-point values. Assume that the velocities are known (or can be interpolated) at the control-volume boundaries. This will be the case, e.g., if you are working on a C-grid and define the concentrations in pressure or water-level points (fig. 11.2). The diffusive part is almost always approximated as

$$\left(D_h \frac{\partial c}{\partial x}\right)_{i+1/2,\,j,k} = D_h \frac{c_{i+1,\,j,k} - c_{i,j,k}}{\Delta x} \tag{11.40}$$

which leads to the familiar 3-point scheme for diffusion terms.

The difficulty is in the advective part of the fluxes. A few possibilities are shown. Central differences are obtained if you take the obvious and simple choice

$$c_{i+1/2,j,k} = \frac{1}{2}\left(c_{i+1,j,k} + c_{i,j,k}\right) \tag{11.41}$$

Although this leads to a second-order accurate approximation of the advection terms, the method is often not very useful, as it is susceptible to numerical oscillations or wiggles. You can show (see any text book on computational fluid dynamics, e.g. Fletcher (1988), Vreugdenhil (1989), Vreugdenhil & Koren (1993)) these to be possible if the *cell-Péclet or cell-Reynolds number* exceeds a certain value (2 in simple cases) so to avoid them you should require:

$$P = \frac{|u|\Delta x}{D} < 2 \tag{11.42}$$

This provides a restriction on the spatial resolution which can often be quite unpleasant. Such wiggles do not necessarily arise if the criterion is violated because a source for them is needed, but if they do, this is a sign that steep gradients in the solution, which often occur near boundaries, are not sufficiently resolved and you consequently have to refine the grid. An example is given in fig. 11.5: for $P = 4$ you get a very inaccurate, wiggly solution which cannot resolve the steep gradient at the downstream side. For $P = 1$, however, the numerical solution is very close to the exact one.

A popular but dangerous methods to avoid such wiggles is using the first-order upwind method in which the advective flux is approximated, depending on the local flow direction, as

$$\begin{aligned} u\, c_{i+1/2,\,j,k} &= u\, c_{i,j,k} \quad \text{if} \quad u > 0 \\ &= u\, c_{i+1,\,j,k} \quad \text{if} \quad u < 0 \end{aligned} \tag{11.43}$$

Several variants exist for multi-dimensional flows. This method is extremely simple and gives always smooth solutions. Moreover, you can show that negative concentrations will not occur.However, this is obtained at the penalty of a first-order numerical error which takes the form of a numerical diffusion term with coefficient $D_{num} = |u| \Delta x /2$. Keeping this small compared to the physical diffusion coefficient leads to

$$\frac{|u| \Delta x}{D} << 2$$

which is even more restrictive than (11.42). Otherwise, you obtain a smooth solution to the wrong problem (one with an artificially high diffusion coefficient). This is also shown in fig. 11.5.

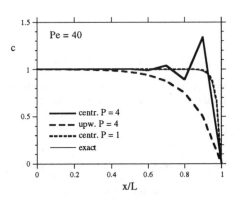

Fig. 11.5. Example of numerical wiggles arising from central differences if cell-Péclet number $P = u\ \Delta x/D$ is too large. Overall Péclet number $Pe = u\ L/D = 40$.

Many modern methods use the upwind idea but with modifications on two important and related issues:
(i) use higher-order interpolation to avoid unphysical diffusion;
(ii) apply some type of limiting such that wiggles do not occur.
The second point is more or less a consequence of the first. As mentioned in section 8.7.3, higher-order methods will not be monotone, i.e. tend to generate oscillations. However, this will usually occur only in small regions. A "smart" finite-difference scheme will recognize this tendency and switch to lower-order methods locally which suppress the wiggles without unduly deteriorating the overall accuracy. As an example (but there are many more methods of this kind), a method by Koren (1993) is shown. The (advective) flux at a cell face is approximated as follows:

$$F_{i+1/2,j,k} = u_{i+1/2}\left\{c_i + \frac{1}{2}\ \phi(r_{i+1/2})(c_i - c_{i-1})\right\} \quad \text{if} \quad u \geq 0$$

$$r_{i+1/2} = \frac{c_{i+1} - c_i + \varepsilon}{c_i - c_{i-1} + \varepsilon}$$

(11.44a)

$$F_{i+1/2,j,k} = u_{i+1/2} \left\{ c_{i+1} + \frac{1}{2} \phi(r_{i+1/2})(c_{i+1} - c_{i+2}) \right\} \quad \text{if} \quad u < 0$$

$$r_{i+1/2} = \frac{c_i - c_{i+1} + \varepsilon}{c_{i+1} - c_{i+2} + \varepsilon}$$

(11.44b)

where on the right-hand side the subscripts j,k on all terms have been suppressed and ε is a small number (say 10^{-10}) to avoid division by zero. The limiter is given by (see fig. 11.6)

$$\begin{aligned}
\phi(r) &= 0 & r &\leq 0 \\
&= 2r & 0 &< r \leq \frac{1}{4} \\
&= (1 + 2r)/3 & \frac{1}{4} &< r \leq 2 \\
&= 2 & r &> 2
\end{aligned}$$

(11.45)

An 1-d example of the results produced by this scheme for a "cloud" of substance advected at constant velocity and diffused by a constant coefficient is shown in fig. 11.7.

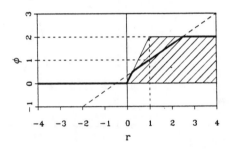

Fig. 11.6. Limiter function in higher-order upwind method (from Koren, 1993)

Stelling (1983) (see also Van Eijkeren et al, 1993) developed what he calls cyclic schemes which can be very well incorporated in ADI methods. For a single dimension with constant advection velocity this has the following appearance:

$$c_i^{n+1/2} = c_i^n - \frac{\Delta t}{24\,\Delta x} u\,(c_{i+2}^n + 4\,c_{i+1}^n - 4\,c_{i-1}^n - c_{i-2}^n)$$

$$c_i^{n+1} = c_i^{n+1/2} - \frac{\Delta t}{4\,\Delta x} u\,(3c_i^{n+1} + 4\,c_{i-1}^{n+1} + c_{i-2}^{n+1})$$

(11.46)

It can be shown that this is third-order accurate in space and second-order in time. The second stage is implicit, but as it contains only upstream values, it can be solved very simply if the flow direction is known. Otherwise some iteration is needed. This method is used in Leendertse's ADI method described in section 11.4. An example of the results for the advection-diffusion eqution is given in fig. 11.7.

Except in the latter case, time has not yet been discretized in the methods discussed in this section. This can again be done in several ways, the most important ones being explicit (forward Euler), implicit (backward Euler or Crank-Nicolson) or ADI. The properties of these are discussed in the general references cited above.

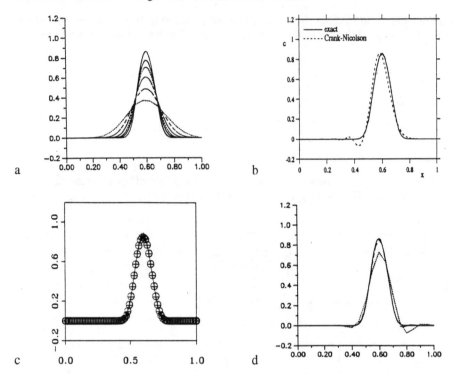

Fig. 11.7. Example of results for one-dimensional advection-diffusion equation (from Vreugdenhil & Koren, 1993). Drawn lines: exact solution. (a) first-order upwind, $\Delta x = 0.05$, 0.025, 0.0125, 0.00625, (b) central differences, $\Delta x = 0.02$, (c) higher-order upwind method with limiter, $\Delta x = 0.0125$, (d) cyclic method, $\Delta x = 0.05$, 0.025

11.6. Accuracy

In two spatial dimensions, the accuracy of numerical methods for the SWE can be assessed by making a number of rather rigorous assumptions, particularly linearizing the equations, assuming coefficients to be constant, forgetting about the boundaries and considering some special limiting cases (see chapters 7 and 9). For the 3-d case, even such a linearized analysis is almost out of question, one of the reasons being that at least the bottom and surface boundaries should be taken into account. So what possibilities do you have to say at least something about numerical accuracy?

1. In some (very) special cases, exact solutions of the 3-d SWE have been constructed, e.g. for a circular island (Sheng, 1987) and small-amplitude oscillating flow over a horizontal bottom (Lynch & Gray, 1985). These give a very useful, but of course only

partial, check on correctness and accuracy of numerical codes (for an example see Blumberg & Herring, 1987).

2. For a horizontal bottom and fixed background stratification, the 3-d flow pattern can be approximately split into vertical modes, each of which is described by the 2-d SWE in the horizontal plane (section 2.5.2). A similar procedure can be followed for the numerical approximation, which reduces the accuracy problem to a 2-d one. This has been done, e.g. by Song & Tang (1993). More generally, you can expect that the results of chapters 7 and 9 apply to each of the modes at least in a qualitative way, provided you interpret depth as the relevant *equivalent depth* (which may be very small for internal modes). The idea is that a numerical method should work well in cases where mode splitting is applicable, even though this does not guarantee that it works in other cases as well.

3. The new aspect is vertical discretization. You could try to envisage a simplified problem which throws light on exactly this aspect. Below, such a problem is analyzed.

4. A general method is numerical convergence analysis. You run a model with increasing spatial and temporal resolution and observe whether the results converge to anything. If so, this will probably be the correct solution and you have information on the rate of convergence. Doing so with different numerical methods in parallel considerably increases insight into their numerical properties. Unfortunately, such exercises are extremely costly and therefore very rarely done. If you double the spatial resolution of a numerical model, the number of grid points increases by a factor of 8. You will have to halve the time step as well (if not worse), so the computational expense increases by a factor of 16 for each grid refinement. Obviously, you will not be able to afford too many steps of refinement.

5. Finally, comparison with measured field data should be mentioned. This is a very common method, which is certainly useful but has its dangers. The problem is that you cannot discern the causes of differences between numerical and observed data, which are almost certain to be found. The causes can be threefold: (i) numerical errors, (ii) wrong numerical values for coefficients such as bottom roughness, viscosity and diffusion coefficients, (iii) measuring errors. The latter two are unavoidable and usually cannot really be controlled. You should avoid, however, adjusting the coefficients to improve agreement between model and observations, if in reality you have insufficient numerical accuracy without knowing it. As a general rule, numerical errors should be an order of magnitude smaller than inaccuracies from the other sources.

As a general and comprehensive analysis of accuracy is lacking, you are best advised to use some combination of the various approaches. Together, they should give you a reasonably complete picture of the performance of your model.

11.6.1. Accuracy of vertical discretization

To get a tractable test problem, neglect advective and Coriolis terms, horizontal viscosity and density effects from (11.3) and assume the free-surface gradient to be a specified function of time:

$$\frac{\partial u}{\partial t} - \frac{\partial}{\partial z}\left(v_t \frac{\partial u}{\partial z}\right) = P\, e^{i\omega t}$$

where P is a typical value of the surface slope. Furthermore, assume a linear eddy-viscosity profile similar to (11.8)

$$v_t = v_1 z$$

This is not very accurate but on the other hand not totally unrealistic either; for steady flow, it produces a logarithmic-linear velocity profile. It contains the typical difficulty of nearly vanishing viscosity near the bottom in turbulent flow. Boundary conditions are

$$u = 0 \quad \text{at} \quad z = z_b + z_0$$
$$\frac{\partial u}{\partial z} = 0 \quad \text{at} \quad z = h$$

This schematized problem is a simplification of one proposed by Davies (1991) who has a somewhat more complicated viscosity profile and includes Coriolis effects. In the present formulation, you can expect a periodic solution in time. Making all variables dimensionless as

$$z' = \frac{z - z_b}{a}, \quad r = \frac{z_0}{a}, \quad u = \frac{a}{v_1} A(z') \, e^{i\omega t}, \quad \sigma = \frac{\omega a}{P v_1}$$

gives the following two-point boundary-value problem for amplitude A (omitting primes)

$$i\sigma A - \frac{d}{dz}\left(z \frac{dA}{dz}\right) = 1$$
$$z = r \quad A = 0 \tag{11.47}$$
$$z = 1 \quad \frac{dA}{dz} = 0$$

The solution can be found analytically:

$$A(z) = -\frac{i}{\sigma} + A_1 J_0\left(\sqrt{2\sigma z}\,(i-1)\right) + A_2 Y_0\left(\sqrt{2\sigma z}\,(i-1)\right) \tag{11.48}$$

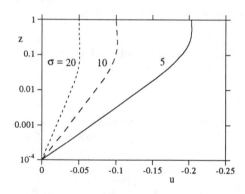

Fig. 11.8. Exact solution of schematic 3-d problem (imaginary part only). Note log scale for z.

where J_0 and Y_0 are Bessel functions and the two constants $A_{1,2}$ are determined from the boundary conditions. The exact solution is illustrated in fig. 11.8 for three values of the dimensionless frequency σ.

This schematic problem is solved numerically by three methods:
1. A finite-difference method with either uniform or expanding grid;
2. A finite-element method with either uniform or expanding grid;
3. Davies' spectral method.
A brief description of each method follows (see also section 11.3).

The finite-difference method on a uniform grid is given by

$$i\sigma A_j - \frac{1}{\Delta z^2} \{z_{j+1/2} (A_{j+1} - A_j) - z_{j-1/2} (A_j - A_{j-1})\} = 1 \qquad (11.49)$$

for j = 2,..., J with

$$A_1 = 0, \qquad A_{J+1} = A_J$$

and the top boundary is halfway between grid point J and $J+1$ for second-order accuracy, so $\Delta z = (1 - r)/(J - 0.5)$. These equations are tridiagonal.

The variant with an expanding grid is as follows.

$$i\sigma A_j - \frac{2}{\Delta z_{j-1} + \Delta z_j}\left(z_{j+1/2} \frac{A_{j+1} - A_j}{\Delta z_j} - z_{j-1/2} \frac{A_j - A_{j-1}}{\Delta z_{j-1}}\right) = 1 \qquad (11.50)$$

where $\Delta z_j = \Delta z_{j-1}(1 + \kappa \Delta z_{j-1})$ and κ can used to adjust the expansion rate. Boundary conditions are as before. For arbitrary distributions of grid points, the truncation error is of first order, but you can show it to be of second order in this special case where the change of grid size is proportional to grid size itself. In order to get the right number of grid points, such that $z_1 = r$ and

$$\tfrac{1}{2}(z_J + z_{J+1}) = 1$$

Table 11.2. Parameters for expanding grid

J	Δz_1 ($\kappa = 1$)	Δz_1 ($\kappa = 2$)	Δz_1 ($\kappa = 4$)	Δz_1 ($\kappa = 8$)
8	0.09143	0.06846	0.04505	0.02676
16	0.04252	0.03054	0.01887	0.01047
32	0.02049	0.01438	0.008559	0.004576
64	0.01006	0.00697	0.004064	0.002116

---

you have to adjust Δz_l for each J and κ. The values shown in table 11.2 have been used in the example below.

The finite-element method is obtained by the standard Galerkin method, using piecewise linear interpolation functions:

$$\sum_{j=k-1}^{k+1} (i\sigma a_{kj}^{(1)} + a_{kj}^{(2)}) A_j = b_k \tag{11.51}$$

where

$$a_{k,\,k-1}^{(1)} = \frac{1}{6}\,(z_k - z_{k-1}) \qquad\qquad a_{k,\,k-1}^{(2)} = \frac{1}{2}\,\frac{z_k + z_{k-1}}{z_k - z_{k-1}}$$

$$a_{k,\,k}^{(1)} = \frac{1}{3}\,(z_{k+1} - z_{k-1}) \qquad\quad a_{k,\,k}^{(2)} = -\,a_{k,\,k-1}^{(2)} - a_{k,\,k+1}^{(2)}$$

$$a_{k,\,k+1}^{(1)} = \frac{1}{6}\,(z_{k+1} - z_k) \qquad\quad a_{k,\,k+1}^{(2)} = \frac{1}{2}\,\frac{z_{k+1} + z_k}{z_{k+1} - z_k}$$

$$b_k = \frac{1}{2}\,(z_{k+1} - z_{k-1})$$

with appropriate modifications at the upper boundary. Either uniformly distributed or other grids may be inserted.

Finally, for the spectral method, you need the eigenfunctions of

$$\frac{d}{dz}\left(z\,\frac{d\phi}{dz}\right) = -\lambda\phi$$

which can be shown to be

$$\phi_j = C_j\,\{J_0(\mu_j\sqrt{z})\,Y_1(\mu_j) - Y_0(\mu_j\sqrt{z})\,J_1(\mu_j)\}$$

where $\mu_j = 2\sqrt{\lambda_j}$ and μ_j are roots of the nonlinear equation

$$J_0(\mu\sqrt{r})\,Y_1(\mu) - Y_0(\mu\sqrt{r})\,J_1(\mu) = 0$$

and the constants C_j are adjusted to normalize the eigenfunctions. Inserting these into (11.31) gives explicit solutions for the coefficients in the eigenfunction expansion:

$$A(z) = \sum_{j=1}^{J} A_j\phi_j(z) \qquad \text{with} \qquad A_j = \frac{\alpha_j}{i\sigma + \lambda_j} \tag{11.52}$$

The first ten eigenvalues are given in Table 11.3.

Numerical results for finite-difference, finite-element and spectral methods are given for the following choice of parameters: $\sigma = 10$, $r = 10^{-4}$. The number of grid intervals J has been varied between 8 and 64.

Table 11.3. Eigenvalues for spectral method

---

n	m
1	0.71669
2	4.28995
3	7.54640
4	10.7663
5	13.9720
6	17.1701
7	20.3634
8	23.5534
9	26.7410
10	29.9268

Fig. 11.9 gives rms. errors for finite-difference and finite-element solutions, either at regular (equidistant) grids or expanding grids with $\kappa = 1$. The conclusion must be that all numerical solutions are relatively poor and that there is only a slow convergence (first-order at most) for increasing J (decreasing Δz). Except at low resolution, the accuracies of finite-difference and FEM methods are equivalent. Fig. 11.10 shows the effect of choosing a more strongly stretched grid (higher expansion coefficient κ). Results are shown only for finite differences, being almost indistinguishable from those for FEM. A more rapidly expanding grid clearly gives a better accuracy. The numerical problem is located in the very strong gradient near the bottom, which requires a very fine grid. Away from the bottom, the solution is fairly smooth and does not require a high resolution. Note that this is not a general conclusion. In oceanographic models, frictional processes are important at the surface and it is customary to have highest resolution there.

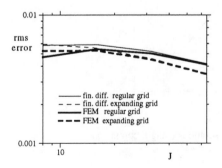

Fig. 11.9. Errors in numerical finite-difference and FEM solutions as functions of number of grid intervals *J*

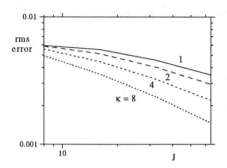

Fig. 11.10. Errors in finite-difference solutions for various values of the expansion coefficient κ

For the spectral method, approximations have been computed using up to 10 spectral components. To compute the integrals involved in the procedure, it proved necessary to apply a logarithmic transformation in order to resolve the steep gradients near the bottom. The rms. errors are shown in fig. 11.11. After a fast convergence until 5 components, the error levels off. This is very probably due to limited accuracy of the numerically computed Bessel functions. Two conclusions are obvious when you compare this with the previous methods: convergence is very much faster and errors are at a much lower level. Davies' (1991) claim that 5 components already give a good accuracy are confirmed. However, reservations on the generality of the spectral method remain valid.

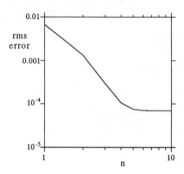

Fig. 11.11. Errors in spectral solutions using *n* components

11.6.2. Hydrostatic errors

A special accuracy problem arises if you use σ-coordinates (section 11.3.2). It is found that spurious pressure gradients arise from the density distribution, even if the latter is horizontally homogeneous, so that no flows should result. The problem has been recognized early in meteorological models, related with flow over mountains (see, e.g. Mesinger, 1982). Several partial solutions have been presented, though apparently none completely satisfactory (e.g. Stelling & Van Kester, 1994).

The problem can be recognized in (11.24). The right-hand side results from $\partial p/\partial x$ in the original Cartesian frame. As shown in (11.3), it can be expressed in a surface-slope part (which does not give any problems) and a part generated by density gradients. If $\partial p/\partial x = 0$, no density-driven flow should result. However, in σ-coordinates this term is split into parts $\partial p/\partial x'$ (σ constant) and $\partial p/\partial \sigma$, neither of which is zero, but which should cancel in this case. In finite-difference form, cancellation is not exact. You get a difference between two large terms, with a small, but possible non-negligible remainder. The steeper the bottom slope, the stronger the coordinate lines are distorted and the larger possible errors are. Fig. 11.12. shows what happens in an exaggerated form. Even if the density profile is independent of x, it will be sampled at different locations (open and black dots) to compute the hydrostatic pressure distributions in two adjacent sections, both shown. Slight horizontal pressure gradients result which act as a driving force. It is possible to show that (for smooth density distributions) the error is proportional to the bottom slope squared and of second order in Δx. In general, therefore, the effect can be avoided by having small bottom slopes (which is required for use of the SWE anyway) and small horizontal grid size.

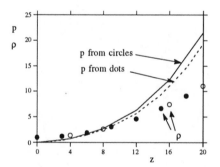

Fig. 11.12. Errors in hydrostatic pressure arising from a horizontally uniform density distribution on a σ-grid

A rather extreme example with a bottom slope of 50% is given by Stelling & Van Kester (1994) and reproduced in fig. 11.13. Density contours are horizontal initially, so no flow should result. The computed flow pattern is completely due to discretization errors. It is, of course questionable whether the shallow-water approximation is valid in this case, but the effect of discretization errors is clearly brought out. Stelling & Van Kester give a method of discretizing the pressure gradient which eliminates a great part (though not all) of the pressure error.

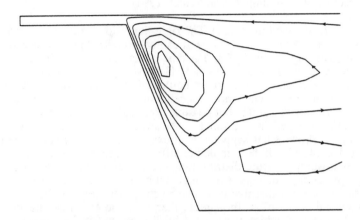

Fig. 11.13. Example of spurious flow generated by errors in hydrostatic pressure
(from Stelling & Van Kester, 1994)

List of notation

In the column "definition", the location is given where a variable is first defined (if applicable). Equation numbers are given in parentheses, e.g. (2.1), section numbers without, e.g. 2.1.

symbol	meaning	definition	dimension
a	depth of flow		m
A	coefficient matrix	2.9	
A	subgrid-scale eddy viscosity	(11.3)	m^2s^{-1}
B	coefficient matrix	2.9	
c	propagation speed		$m\,s^{-1}$
c_f	friction coefficient	(2.42)	-
c_g	group velocity		$m\,s^{-1}$
c_r	relative propagation speed	(7.4)	-
d	numerical damping factor	(7.3)	-
D	diffusion coefficient	(2.4)	m^2s^{-1}
D_h, D_v	eddy diffusion coefficients	(11.6)	m^2s^{-1}
D_{0x}	central difference operator	(6.1)	m^{-1}
D_{+x}	forward difference operator	(6.2)	m^{-1}
f	Coriolis parameter	(2.1)	s^{-1}
F	Froude number	2.1	-
g	acceleration of gravity		$m\,s^{-2}$
h	surface level	(2.9)	m
k	wave number		m^{-1}
k	turbulent kinetic energy	(11.11)	m^2s^{-2}
l	turbulence length scale	11.2.2	m
M	mass matrix	6.4	
M_x, M_y, M_l	averaging operators	6.2	-
n	normal coordinate		m
p	pressure		$N\,m^{-2}$
p_a	atmospheric pressure	(2.11)	$N\,m^{-2}$
P	cell Péclet number	(11.42)	-
Pe	Péclet number	11.5	-
q	potential vorticity	3.3.4	$m^{-1}s^{-1}$
r	linearized friction coefficient	(4.2)	s^{-1}
R	Rossby deformation radius	4.4.1	m
R	reflection coefficient	(5.21)	-
Re	Reynolds number	2.3	-
Ri	Richardson number	11.2.2	-
Ro	Rossby number	2.3	-

s	tangential coordinate		m
S	salinity	2.1	-
t	time		s
T	temperature		K
T_{ij}	lateral momentum fluxes	(2.19)	N m^{-2}
u	velocity in x direction		m s^{-1}
U	"frozen" advection velocity	4.1	m s^{-1}
v	velocity in y direction		m s^{-1}
V	"frozen" advection velocity	4.1	m s^{-1}
$\mathbf{v}$	vector of unknowns	2.9	
w	velocity in z direction		m s^{-1}
w	work per time step and grid point	9.7	-
W	amount of work	9.7	-
x	coordinate		m
y	coordinate		m
z	vertical coordinate		m
z_b	bottom level	(2.8)	m
z_0	measure of bottom roughness	11.2.2	m
β	gradient of Coriolis parameter	4.1	m^{-1}s^{-1}
Δt	time step	6.1	s
$\Delta x, \Delta y$	grid sizes	6.1	m
ε	relative density difference	2.5.3	-
ε	turbulent energy dissipation rate	2.8.2	m s^{-2}
ζ	vorticity	(3.6)	s^{-1}
η	curvilinear coordinate	(2.53)	m
η	ratio of grid size to wave length	7.1	-
θ	implicit weighting factor	(8.10)	-
κ	Von Karman's constant	2.8.1	-
ν	viscosity	(2.2)	m^{2}s^{-1}
ν	frequency	(7.1)	s^{-1}
ν_t	turbulent or eddy viscosity		m^{2}s^{-1}
ξ	curvilinear coordinate	(2.53)	m
ξ	ratio of grid size to wave length	7.1	-
ρ	density	(2.5)	kg m^{-3}
ρ	amplification factor	(9.1)	-
σ	Courant number	(8.5)	-
σ	scaled vertical coordinate	(11.22)	-
τ	ratio of time step to wave period	9.2	-
τ_{ij}	viscous stress	(2.2)	N m^{-2}
τ_{bx}, τ_{by}	bottom stress		N m^{-2}
ϕ	interpolation function	6.4	-
ψ	stream function	(6.46)	m^{2}s^{-1}
ω	frequency		s^{-1}

References

Abbott, M.B., A. Damsgaard & G.S. Rodenhuis 1972 - System 21 Jupiter (A design system for two-dimensional nearly-horizontal flows), J. Hydr. Res. **11**, 1, 1-28

Alcrudo, F. & P. Garcia-Navarro 1993 - A high-resolution Godunov-type scheme in finite volumes for the 2-d shallow-water equations, Int. J. Num. Meth. Fluids **16**, 489-505

Arakawa, A. 1966 - Computational design for long-term numerical integration of the equations for fluid motion: two-dimensional incompressible flow, J. Comp. Phys. **1**, 119-143

Arakawa, A. & V.R. Lamb 1981 - A potential enstrophy and energy conserving scheme for the shallow-water equations, Mon. Wea. Rev. **109**, 18-36

Arakawa, A. & Y-J Hsu 1990 - Energy conserving and potential-enstrophy dissipating schemes for the shallow-water equations, Mon. Wea. Rev. **118**, 1960-1969

Arina, R. 1986 - Orthogonal grids with adaptive control, in: (J. Hauser & C. Taylor, eds.) Proc. Int. Conf. Numerical Grid Generation in Comp. Fluid Dyn., Pineridge Press, 113-124

Aris, R. 1962 - Vectors, tensors and the basic equations of fluid mechanics, Prentice-Hall

Batchelor, G.K. 1967 - An introduction to fluid dynamics, Cambridge Univ. Press

Battjes, J.J. 1988 - Surfzone dynamics, Ann. Rev. Fluid Mech. **20**, 257-293

Beam, R.M. & R.F. Warming 1976 - An implicit finite-difference algorithm for hyperbolic equations in conservation-law form, J. Comp. Phys. **22**, 87-110

Beam, R.M. & R.F. Warming 1978 - An implicit factored scheme for the compressible Navier-Stokes equations, AIAA J. **16**, 4, 393-402

Beam, R.M., R.F. Warming & H.C. Yee 1982 - Stability analysis of numerical boundary conditions and implicit difference approximations for hyperbolic equations, J. Comp. Phys. **48**, 200-222

Benqué, J.P., J.A. Cunge, J. Feuillet, A. Hauguel & F.M. Holly 1982 - New method for tidal current computation, J. Waterway, Port, Coastal & Ocean Div. ASCE, **108**, WW3, 396-417

Blumberg, A.F. & H.J. Herring 1987 - Circulation modelling using orthogonal curvilinear cordinates, in: Nihoul & Jamart (1987) 55-88

Borthwick, A.G.L. & R.W. Barber 1992 - River and reservoir flow modelling using the transformed shallow-water equations, Int. J. Num. Meth. in Fluids, **14**, 1193-1217

Briley, W.R. & H. McDonald, 1980 - On the structure and use of linearized block implicit schemes, J. Comp. Phys. **34**, 54-73

Bryan, K. 1969 - A numerical method for the study of the circulation of the world ocean, J. Comp. Phys. **4**, 347-376

Canuto, C., M.Y. Hussaini, A. Quarteroni & T.A. Zang 1988 - Spectral methods in fluid dynamics, Springer, New York

Charney, J.G., R. Fjörtoft & J. von Neumann 1950 - Numerical integration of the barotropic vorticity equation, Tellus **2** ,237-254

Chock, D.P. & A.M. Dunker 1983/1985/1991 - A comparison of numerical methods for solving the advection equation, Atmosph. Environment (part 1) **17**, 11-24, (part 2) **19**, 571-586, (part 3) **25A**, 853-871

Courant, R. & K.O. Friedrichs 1948 - Supersonic flow and shock waves, Interscience Publ. New York

Courant, R. & D. Hilbert 1962 - Methods of mathematical physics, Interscience Publ. New York

Davies, A.M. 1987 - Spectral models in continental shelf sea oceanography, in: Heaps (1987), 71-106

Davies, A.M. 1991 - On the accuracy of finite difference and modal methods for computing tidal and wind wave current profiles, Int. J. Numer. Meth. in Fluids **12**, 101-124

Davies, A.M. & H. Gerritsen 1994 - An intercomparison of three-dimensional tidal hydrodynamic models of the Irish Sea, Tellus **46A**, 200-221

Deardorff, J.W. 1971 - On the magnitude of the subgrid-scale eddy coefficient, J. Comp. Phys. **7**,1, 120-133

Douglas, J. & J.E. Gunn, 1964 - A general formulation of alternating-direction implicit methods, Num. Math. **6**, 428

Douglas, J. & H.H. Rachford, 1958 - On the numerical solution of heat-conduction problems in two and three space variables, Trans. Am. Math. Soc. **82**, 421-439

Dowling, T.E. & A.P. Ingersoll, 1989 - Jupiter's Great Red Spot as a shallow-water system, J. Atmosph. Sci. **46**, 21, 3256-3278

Dronkers, J.J. 1964 - Tidal computations in rivers and coastal waters, North-Holland Publ., Amsterdam

Dube, S.K., P.C. Sinha & G.D. Roy 1985 - The numerical simulation of storm surges along the Bangla Desh coast, Dyn. Atm. Oceans **9**, 121-133

Eiseman, P.R. 1988 - Grid generation for fluid mechanics computations, Ann. Rev. Fluid Mech. **17**, 487-522

Elvius, T. & A. Sundström 1973 - Computationally efficient schemes and boundary conditions for a fine-mesh barotropic model based on the shallow-water equations, Tellus **25**, 2, 132-156

Engquist, B. & A. Majda 1977 - Absorbing boundary conditions for the numerical simulation of waves, Math. Comp. **31**, 629-651

Fairweather, G. & I.M. Navon, 1980 - A linear ADI method for the shallow-water equations, J. Comp. Phys. **37**, 1-18

Fischer, G. 1959 - Ein numerisches Verfahren zur Errechnung van Windstau und Gezeiten in Randmeeren, Tellus **11**,1, 1-18

Fischer, H.B., E.J. List & R.C.Y. Kon 1979 - Mixing in inland and coastal waters, Academic Press

Fletcher, C.A.J. 1984 - Computational Galerkin methods, Springer, Berlin

Fletcher, C.A.J. 1988 - Computational techniques for fluid dynamics, Springer Verlag, 2 volumes

Flokstra, C. 1977 - The closure problem for depth-averaged two-dimensional flow, 17th Int. Congress IAHR, Baden-Baden, paper A106

Gadd, A.J. 1978 - A split explicit integration scheme for numerical weather prediction, Quart. J. Roy. Met. Soc. **104**, 569-582

Garvine, R.W. 1987 - Estuary plume and fronts in shelf waters: a layer model, J. Phys. Oceanogr. 17, 1877-1896

Gerritsen, H. 1985 - Some first particle track computations with fine grid and coarse grid velocities in the Southern North Sea, Delft Hydraulics, Publ. 351

Gill, A.E. 1982 - Atmosphere-Ocean Dynamics, Acad. Press

Glaister, P. 1993 - Flux-difference splitting for open-channel flows, Int. J. Num. Meth. Fluids **16**, 629-654

Gopalakrishnan, T.C. 1989 - A moving boundary circulation model for regions with large tidal flats, Int. J. Num. Meth. in Eng. **28**, 2, 245-260

Groen, P. & G.W. Groves 1962 - Surges, in: (M.N. Hill, ed) The Sea , vol I, Interscience, p. 611-646

Gustafsson, B. 1971 - An alternating-direction implicit method for solving the shallow-water equations, J. Comp. Phys. **7**, 239-254

Gustafsson, B. 1975 - The convergence rate for difference approximations to mixed initial-boundary value problems, Math. Comp. **29**, 396-406

Gustafsson, B., H.O. Kreiss & A. Sundström 1972 - Stability theory of difference approximations for mixed initial-boundary value problems, Math. Comp. **26**, 649-686

Haidvogel, D.B., J.L. Wilkin & R. Young 1991 - A semi-spectral primitive equation ocean circulation model using vertical sigma and orthogonal horizontal coordinates, J. Comp. Phys. **94**, 151-185

Haltiner, G.J. & R.T. Williams, 1980 - Numerical prediction and dynamic meteorology, Wiley, New York

Hansen,W. 1956 - Theorie zur Errechnung des Wasserstandes und der Strömungen in Randmeeren nebst Anwendungen, Tellus **8**,3, 187-300

Häuser, J. & C. Taylor, eds. 1986 - Proc. Int. Conf. Numerical Grid Generation in Comp. Fluid Dyn., Pineridge Press

Häuser, J., H.G. Paap, D. Eppel & A. Müller, 1985 - Solution of shallow-water equations for complex flow domains via boundary-fitted coordinates, Int. J. Num. Meth. in Fluids, **5**, 727-744

Heaps, N.S. 1967 - Storm surges, Ocean. Mar. Biol. Ann. Rev. **5**, 11-47

Heaps, N.S. 1973 - Three-dimensional numerical model of the Irish Sea, Geophys. J. Roy. Astron. Soc. **35**, 99-120

Heaps, N.S. (ed.) 1987 - Three-dimensional coastal ocean models, Coastal and Estuarine Sciences vol. 4, Amer. Geophys. Union, Washington, D.C.

Hendershott, M.C. 1981 - Long waves and ocean tides, Ch. 10 in: B.A. Warren & C. Wunsch (eds.) Evolution of physical oceanography, MIT Press, 292-341

Higdon, R.L. 1986 - Absorbing boundary conditions for difference approximations to the multi-dimensional wave equation, Math. Comp. **47**, 176, 437-459

Higdon, R.L. 1987 - Absorbing boundary conditions for the wave equation, Math. Comp. **49**, 179, 65-90

Hinze, J.O. 1975 - Turbulence (2nd ed.), McGraw- Hill, New York

Hirsch, C. 1988, 1990 - Numerical computation of internal and external flows, Wiley, London (2 vols.)

Holsters, H. 1961 - Remarques sur la stabilité dans les calculs de marée, Proc. Symp. Math.-Hydrodyn. Meth. in Phys. Oceanogr., Hamburg, 211-225

Janjic, Z.I. 1984 - Nonlinear advection schemes and energy cascade on semi-staggered grids, Mon. Wea. Rev. **112**, 1234-1245

Janjic, Z.I. & F. Mesinger 1983 - Finite-difference methods for the shallow water equations in various horizontal grids, Seminar "Numerical methods for Weather Prediction", European Centre for Medium Range Weather Forecasts, vol. I, 29-101

Jarraud, M. & A.J. Simmons 1983 - The spectral technique, in ECMWF Seminar Numerical Methods for Weather Prediction, vol.2, 1- 59

Jarraud, M. & A.P.M. Baede 1985 - The use of spectral techniques in numerical weather prediction, Am. Math. Soc., Lectures in Appl. Math. 22, 1 -41

Johns, B., S.K. Dube, P.C. Sinha, U.C. Mohanty & A.D. Rao 1982 - The simulation of continuously deforming lateral boundaries in problems involving the shallow-water equations, Comp. & Fluids **10**, 2, 105-116

Johns, B. & T. Oguz 1987 - Turbulent energy closure schemes, in: Heaps (1987), 17-40

Johnson, B.H. 1982 - Numerical modelling of estuarine hydrodynamics on a boundary fitted coordinate system, in: J.F. Thompson (ed.) Numerical grid generation, Elsevier, 409-436

Johnson, B.H. & J.F Thompson 1986 - Discussion of a depth-dependent adaptive grid generator for use in computational hydraulics, in: (J. Häuser & C. Taylor, eds.) Proc. Int. Conf. Numerical Grid Generation in Comp. Fluid Dyn., Pineridge Press, 629-640

Kalkwijk, J.P.T. & H.J. de Vriend 1980 - Computation of the flow in shallow river bends, J. Hydr. Res. **18**, 4, 327-342

Kinnmark, I. 1986 - The shallow-water wave equations: formulation, analysis and application, Springer, Berlin

Kowalik, Z. & T.S. Murty 1993 - Numerical modelling of ocean dynamics, World Scientific Publ., Singapore

Kreiss, H. & J. Oliger 1973 - Methods for the approximate solution of time dependent problems, GARP Publ. Ser. 10, World Met. Org.

Kuipers, J. & C.B. Vreugdenhil 1973 - Calculations of two-dimensional horizontal flow, Delft Hydr. Lab. Report S 163 - I

Kutler, P. (ed.) 1982 - Numerical boundary condition procedures, NASA Conf. Publ. 201

Kwizak, M. & A. Roberts 1971 - Implicit integration of a grid point model, Mon. Weather Rev. **99**, 32-36

Lamb, Sir H. 1932 - Hydrodynamics, Dover, New York

Launder, B.E. 1989 - The prediction of force field effects on turbulent shear flows via second-moment closure, in: H.H. Fernholz & H.E. Fiedler - Advances in Turbulence 2, Springer, Heidelberg, pp 338-358

Lax, P.D. & B. Wendroff 1960 - Systems of conservation laws, Comm. Pure & Appl. Math. **13**, 217-237

Lean, G.H. & T.J. Weare, 1979 - Modeling two-dimensional circulating flow, J. Hydr. Div. ASCE **105**, HY1, 17-26

Leendertse, J.J. 1967 - Aspects of a computational model for long-period water-wave propagation, RAND Corp. Memorandum RM-5294-PR

Leendertse, J.J. 1989 - A new approach to three-dimensional free-surface flow modelling, RAND Corp. R-3712-NETH/RC

Leendertse, J.J., R.C. Alexander & S-K Liu 1973 - A three-dimensional model for estuaries and coastal seas. Vol. I, Principles of computation, RAND Corp. R-1417-OWRR

Le Méhauté, B. 1976 - An introduction to hydrodynamics and water waves, Springer, New York

LeVeque, R.J. 1990 - Numerical methods for conservation laws, Birkhäuser, Basel

Longuet-Higgins, M.S. & R.W. Stewart 1964 - Radiation stress in water waves, a physical discussion with applications, Deep-Sea Res. **11**, 529-562

Lorentz, H.A. 1937 - Accounting for friction for oscillating fluid motions (in Dutch), De Ingenieur, 695, also in Coll. Papers, 4, 252, Nijhoff, Den Haag.

Lumley, J.L. & H.A. Panofsky 1964 - The structure of atmospheric turbulence, Interscience Publ.

Lynch, D.R. & W.G. Gray 1979 - A wave equation model for finite element tidal computations, Comp. Fluids 7,3, 207-228

Lynch, D.R. & C.B. Officer 1985 - Analytic test cases for three-dimensional hydrodynamic models, Int. J. Num. Meth. in Fluids **5**, 529-543

Maday, Y., A.T. Patera & E.M. Rønquist 1990 - An operator-integration-factor splitting method for time-dependent problems: application to imcompressible fluid flow, J. Scient. Comp. **5**, 263-292

McGuirk, J. & W. Rodi 1978 - A depth-averaged model for side discharges into open-channel flow, J. Fluid Mech. **86**, 4, 761-782

Mellor, G.J. & A.F. Blumberg 1985 - Modeling vertical and horizontal diffusivities with the sigma coordinate system, J. Phys. Oceanogr. **113**, 1379-1383

Mellor, G.J. & T. Yamada 1982 - Development of a turbulence closure model for geophysical fluid problems, Rev. Geophys. & Space Phys. **20**, 851-875

Mesinger, F. 1982 - On the convergence and error problems of the calculation of the pressure gradient in sigma coordinate models, Geophys. Astrophys. Fluid Dyn. **19**, 105-117

Mynett, A.E., P. Wesseling, A. Segal & C.G.M. Kassels 1991 - The Isnas incompressible Navier-Stokes solver: invariant discretization, Appl. Sci. Res. **48**,2, 175-191

Niemeyer, G. 1979 - Efficient simulation of nonlinear steady flow, J. Hydr. Div. ASCE **105**, 185-195

Nihoul, J.C.J. & F.C. Ronday 1975 - The influence of tidal stress on the residual circulation, Tellus **29**, 484-490

Nihoul, J.C.J. & B.M. Jamart (eds.) 1987 - Three-dimensional models of marine and estuarine dynamics, Elsevier Oceanography Series vol. 45, Elsevier, Amsterdam

Officier, M.J., C.B. Vreugdenhil & H.G. Wind, 1986 - Applications in hydraulics of numerical solutions of the Navier-Stokes equations, in: C. Taylor et al (eds.) Computational techniques for fluid flow, Pineridge Press, Swansea, p.115-147

Ogink, H.J.M. 1985 - On the effective viscosity coefficient in 2-D depth-averaged flow models, 21st Congress IAHR, Melbourne, vol.3 , 474-479

Ogink, H.J.M., J.G. Grijsen & A.J.H. Wijbenga 1986 - Aspects of flood level computations, Int. Symp. Flood Frequency and Risk Analysis, Baton Rouge, USA (also Delft Hydraulics Comm. 357)

Oliger, J. & A. Sundström 1978 - Theoretical and practical aspects of some initial boundary value problems in fluid dynamics, SIAM J. **35**, 419-446

Peaceman, D.W. & H.H. Rachford 1955 - The numerical solution of parabolic and elliptic differential equations, SIAM J. **3**, 28-41

Pedlosky, J. 1979 - Geophysical fluid dynamics, Springer, New York

Perrels, P.A.J. & M. Karelse 1981 - A two-dimensional, laterally averaged model for salt intrusion in estuaries, in: Transport models for inland and coastal waters, Academic Press, 483-535

Peyret, R. & T.D. Taylor 1983 - Computational methods for fluid flow, Springer, New York

Platzman, G.W. 1972 - Two-dimensional free oscillations in natural basins, J. Phys. Oceanogr. **2**, 2, 117-138

Phillips, N.A. 1957 - A coordinate system having some special advantages for numerical forecasting, J. Meteorol. **14**, 184-185

Phillips, O.M. 1977 - The dynamics of the upper ocean, Cambridge Univ. Press, Cambridge

Praagman, N. 1979 - Numerical solution of the shallow-water equations, Ph. D. thesis Delft Univ. of Techn.

Priestley, A. 1987 - The use of a characteristics based scheme for the 2-D shallow-water equations, ECMWF Workshop Techniques for horizontal discretization in numerical weather prediction models, 157-186

Priestley, A. 1993 - A quasi-Riemannian method for the solution of one-dimensional shallow-water flow, J. Comp. Phys. **106**, 139-146

Rastogi, A.K. & W. Rodi 1978 - Predictions of heat and mass transfer in open channels, J. Hydr. Div. ASCE HY3, 397-420

Reid, R.O. 1957 - Modification of the quadratic bottom stress for turbulent channel flow in the presence of surface wind-stress, US Army Corps of Engineers, Beach Erosion Board, Tech. Mem. no 93

Ridderinkhof, H. 1990 - Residual currents and mixing in the Wadden Sea, PhD thesis, Univ. of Utrecht

Robert, J.L. & Y. Ouellet 1987 - A three-dimensional finite-element model for the study of steady and non-steady natural flows, in: Nihoul & Jamart (1987), 359-372

Robinson, I.S. 1983 - Tidally induced residual flows, Ch.7 in: B. Johns (ed.) Physical Oceanography of coastal and shelf seas, Elsevier Publ. Amsterdam, p. 321-356

Rodi, W. 1980 - Turbulence models and their application in hydraulics, IAHR, Delft

Roe, P.L. 1981 - Approximate Riemann solvers, parameter vectors and difference schemes, J. Comp. Phys. **43**, 357-372

Roe, P.L. 1987 - Upwind differencing schemes for hyperbolic conservation laws with source terms, in: (C. Carasso et al, eds.) - Nonlinear hyperbolic problems, Springer, Berlin 1987, pp. 41-51

Roe, P.L. 1991 - Discontinuous solutions to hyperbolic systems under operator splitting, in: Num. Meth. for Part. Diff. Eqs. **7**, 277-297

Røed, L.P. & C.K. Cooper 1987 - A study of various open boundary conditions for wind-forced barotropic numerical ocean models, in: Nihoul & Jamart, 1987, 305 - 336

Rood, R.R. 1987 - Numerical advection algorithms and their role in atmospheric transport and chemistry models, Revs. Geophys. **25**, 71-100

Rubesin, M.W. 1977 - Numerical turbulence modelling, AGARD Lecrure Series no. 86, Computational Fluid Dynamics

Rumsey, C.L., B. van Leer & P.L. Roe 1993 - A multidimensional flux function with application to the Euler and Navier-Stokes equations, J. Comp. Phys. **105**, 306-323

Sadourny, R. 1975 - The dynamics of finite-difference models of the shallow-water equations, J. Atmosph. Sci. **32**, 680-689

Sadourny, R. & C. Basdevant 1985 - Parameterization of subgrid scale barotropic and baroclinic eddies in quasi-geostrophic models: anticipated potential vorticity method, J. Atmosph. Sci. **42**, 1353-1363

Salmon, R. & L.D. Talley 1989 - Generalizations of Arakawa's Jacobian, J. Comp. Phys. **83**, 247-259

Schönstadt, A.L. 1977 - The effect of spatial discretization on the steady-state and transient solutions of a dispersive wave equation, J. Comp. Phys. **23**, 364- 379

Sheng, Y.P. 1987 - On modelling three-dimensional estuarine and marine hydrodynamics, in: Nihoul & Jamart (1987), 35-54

Shokin, Y.I. & L.B. Chubarow 1980 - Finite-difference simulation of tsunami propagation, in (U. Müller et al, eds.) Theoretical and experimental fluid mechanics, Springer, Berlin, 599-606

Sielecki, A. 1968 - An energy-conserving difference scheme for the storm-surge equations, Mon. Weather Rev. **96**, 3

Sielecki, A. & M.G. Wurtele 1970 - The numerical integration of the nonlinear shallow-water equations with sloping boundaries, J. Comp. Phys. **6**, 219-236

Simons, T.J. 1980 - Circulation models of lakes and inland seas, Bull. 203, Canadian Dept. of Fisheries and Oceans

Smagorinsky, J. 1963 - General circulation experiments with the primitive equations, Mon. Weather Rev. **91**, 99-164

Song, Y. & T. Tang 1993 - Dispersion and group velocity in numerical schemes for three-dimensional hydrodynamic equations, J. Comp. Phys. **105**, 72-82

Stelling, G.S. 1983 - On the construction of computational methods for shallow-water flow problems, PhD thesis Delft Univ. of Technology

Stelling, G.S., A.K. Wiersma & J.B.T.M. Willemse 1986 - Practical aspects of accurate tidal computations, J. Hydr. Eng. ASCE **112**, 802-817

Stelling, G.S. & J.A.T.M. van Kester 1994 - On the approximation of horizontal gradients in sigma coordinates for bathymetry with steep bottom slopes, Int. J. Numer. Methods in Fluids **18**, 915-935

Strang, G. 1968 - On the construction and comparison of difference schemes, SIAM J. Num. Anal. **5**, 506-517

Strikwerda, J.C. 1989 - Finite difference schemes and partial differential equations, Wadsworth & Brooks/Cole, Pacific Grove, Cal.

Sündermann, J. 1971 - Die hydrodynamisch-numerische Berechnung der Vertikal-struktur von Bewegungsvorgängen in Kanälen und Becken, Mitt. Inst. f. Meereskunde, Univ. Hamburg, XIX

Sundström, A 1977 - Boundary conditions for limited-area integration of the viscous forecast equations, Beitr. zur Physik der Atmosphäre **50**, 218-224

Tan Weiyan 1992 - Shallow-water hydrodynamics, Elseviers Oceanographic Series no. 55, Beijing, Water & Power Press

Taylor, G.I. 1919 - Tidal friction in the Irish Sea, Phil. Trans. Royal Soc. London, 1-33

Thompson, J.F., Z.U.A. Warsi 1982 - Boundary-fitted coordinate systems for numerical solution of partial differential equations - a review, J. Comp. Phys. **47**, 1-108

Thompson, J.F., Z.U.A. Warsi & C.W. Mastin 1985 - Numerical grid generation, foundation and applications, North-Holland, Amsterdam

Trefethen, L.N. 1982 - Group velocity of finite-difference schemes, SIAM Rev. **23**, 113-136

Trefethen, L.N. 1983 - Group velocity interpretation of the stability theory of Gustafsson, Kreiss and Sundström, J. Comp. Phys. **49**, 199-217

Trefethen, L.N. 1985 - Stability of finite-difference models with one or two boundaries, Lect. in Appl. Math. 22, 311-326

Usseglio-Polatera, J.M. & P. Sauvaget 1987 - A coupled 2-d/3-d modelling system for computation of tidal and wind-induced currents, in: Nihoul & Jamart (1987) 539-554

Van der Houwen, P.J. 1977 - Computation of water levels in seas and rivers (in Dutch), Centre for Math., Amsterdam, Syllabus 33

Verboom, G.K. & A. Slob 1984 - Weakly-reflective boundary conditions for two-dimensional shallow water flow problems, Adv. Water Resources 7, 192-197

Verboom, G.K., G.S. Stelling & M.J. Officier 1982 - Boundary conditions for the shallow-water equations, in: M.B. Abbott & J.A. Cunge (eds.) Engineering applications of computational hydraulics, Pitman, London, 230-262

Versteegh, J. 1990 - The numerical simulation of three-dimensional flow through and around hydraulic structures, PhD Thesis Delft Univ. of Technology

Vichnevetsky, R. 1987 - Wave propagation analysis of difference schemes for hyperbolic equations: a review, Int. J. Num. Meth. in Fluids 7, 409-452

Vichnevetsky, R. & J.B. Bowles 1982 - Fourier analysis of numerical approximations of hyperbolic equations, SIAM, Philadelphia

Vreugdenhil, C.B. 1977 - The formulation of wind influence in shallow water, Delft Hydraulics Lab. Report S114 - V

Vreugdenhil, C.B. 1978 - Two-layer flow in two horizontal dimensions, Delft Hydraulics Lab. Report S114 - VI

Vreugdenhil, C.B. 1979 - Two-layer shallow-water flow in two dimensions, a numerical study, J. Comp. Phys. 33, 2, 169-184

Vreugdenhil, C.B. 1989 - Computational hydraulics, Springer , Berlin

Vreugdenhil, C.B. 1990 - Numerical methods for shallow-water flow, Lecture Series on Comp. Fluid Dyn., Von Karman Inst., Rhode-St. Genese, Belgium

Vreugdenhil, C.B. 1991 - Grid control for non-orthogonal elliptic grids, Comm. Appl. Num. Meth. 7, 633-637

Vreugdenhil, C.B. & N. Booij 1985 - Numerical outflow boundary conditions for the shallow-water equations, Int. J. Num. Methods in Fluids 5, 393-397

Vreugdenhil, C.B. & B. Koren (eds.) 1993 - Numerical methods for advection-diffusion problems, Notes on Numerical Fluid Mechanics, vol. 45, Vieweg, Braunschweig

Vreugdenhil, C.B. & J.H.A. Wijbenga 1982 - Computation of flow patterns in rivers, J. Hydr. Div. ASCE 108, HY11, 1296-1310

Weare, T.J. 1979 - Errors arising from irregular boundaries in ADI solutions of the shallow-water equations, Int. J. Num. Methods in Eng. 14, 921-931

Wijbenga, J.H.A. 1985 - Determination of flow patterns in rivers with curvilinear coordinates, in: Proc. 21st Congress IAHR, Melbourne

Wilders, P., T.L. van Stijn, G.S. Stelling & G.A. Fokkema 1988 - A fully implicit splitting method for accurate tidal computations, Int. J. Num. Meth. Eng. **26**, 12, 2707-2721

Willemse, J.B.T.M., G.S. Stelling & G.K. Verboom 1985 - Solving the shallow-water equations with an orthogonal coordinate transformation, Int. Symp. Comp. Fluid Dyn., Tokyo

Wind, H.G. & C.B. Vreugdenhil 1986 - Rip-current generation near structures, J. Fluid Mech. **171**, 459-476

Wubs, F.W. 1987 - Numerical solution of the shallow-water equations, CWI Centre for Math. and Computer Sci. Amsterdam, Tract 49

Zielke, W. 1968 - Frequency dependent friction in transient pipe flow, J. Basic Eng. **D90**, 109-115

Zimmerman, J.T.F. 1978 - Topographic generation of residual circulation by oscillatory tidal currents, Geoph. Astroph. Fluid Dyn. **11**, 35-47

Zimmerman, J.T.F. 1979 - On the Euler-Lagrange transformation and the Stokes drift in the presence of oscillatory and residual currents, Deep Sea Res. **26**, 505 - 520

Zimmerman, J.T.F. 1992 - On the Lorentz linearization of a nonlinearly damped Helmholtz oscillator, Proc. Royal Dutch Acad. Science **95**, 1, 127-145

Index

Water Science and Technology Library

1. A.S. Eikum and R.W. Seabloom (eds.): *Alternative Wastewater Treatment. Low-Cost Small Systems, Research and Development. Proceedings of the Conference held in Oslo, Norway (7–10 September 1981).* 1982
ISBN 90-277-1430-4
2. W. Brutsaert and G.H. Jirka (eds.): *Gas Transfer at Water Surfaces.* 1984
ISBN 90-277-1697-8
3. D.A. Kraijenhoff and J.R. Moll (eds.): *River Flow Modelling and Forecasting.* 1986
ISBN 90-277-2082-7
4. World Meteorological Organization (ed.): *Microprocessors in Operational Hydrology.* Proceedings of a Conference held in Geneva (4–5 September 1984). 1986
ISBN 90-277-2156-4
5. J. Němec: *Hydrological Forecasting.* Design and Operation of Hydrological Forecasting Systems. 1986
ISBN 90-277-2259-5
6. V.K. Gupta, I. Rodríguez-Iturbe and E.F. Wood (eds.): *Scale Problems in Hydrology.* Runoff Generation and Basin Response. 1986
ISBN 90-277-2258-7
7. D.C. Major and H.E. Schwarz: *Large-Scale Regional Water Resources Planning.* The North Atlantic Regional Study. 1990
ISBN 0-7923-0711-9
8. W.H. Hager: *Energy Dissipators and Hydraulic Jump.* 1992
ISBN 0-7923-1508-1
9. V.P. Singh and M. Fiorentino (eds.): *Entropy and Energy Dissipation in Water Resources.* 1992
ISBN 0-7923-1696-7
10. K.W. Hipel (ed.): *Stochastic and Statistical Methods in Hydrology and Environmental Engineering.* A Four Volume Work Resulting from the International Conference in Honour of Professor T. E. Unny (21–23 June 1993). 1994
10/1: Extreme values: floods and droughts
ISBN 0-7923-2756-X
10/2: Stochastic and statistical modelling with groundwater and surface water applications
ISBN 0-7923-2757-8
10/3: Time series analysis in hydrology and environmental engineering
ISBN 0-7923-2758-6
10/4: Effective environmental management for sustainable development
ISBN 0-7923-2759-4
Set 10/1–10/4: ISBN 0-7923-2760-8
11. S. N. Rodionov: *Global and Regional Climate Interaction: The Caspian Sea Experience.* 1994
ISBN 0-7923-2784-5
12. A. Peters, G. Wittum, B. Herrling, U. Meissner, C.A. Brebbia, W.G. Gray and G.F. Pinder (eds.): *Computational Methods in Water Resources X.* 1994
Set 12/1–12/2: ISBN 0-7923-2937-6
13. C. B. Vreugdenhil: *Numerical Methods for Shallow-Water Flow.* 1994
ISBN 0-7923-3164-8

Kluwer Academic Publishers – Dordrecht / Boston / London